AF356460

Surface and Interface Engineering for Organic Device Applications

Editor

Ju-Hyung Kim

MDPI • Basel • Beijing • Wuhan • Barcelona • Belgrade • Manchester • Tokyo • Cluj • Tianjin

Editor
Ju-Hyung Kim
Ajou University
Korea

Editorial Office
MDPI
St. Alban-Anlage 66
4052 Basel, Switzerland

This is a reprint of articles from the Special Issue published online in the open access journal *Materials* (ISSN 1996-1944) (available at: https://www.mdpi.com/journal/materials/special_issues/surf_interface_organic).

For citation purposes, cite each article independently as indicated on the article page online and as indicated below:

LastName, A.A.; LastName, B.B.; LastName, C.C. Article Title. *Journal Name* **Year**, *Volume Number*, Page Range.

ISBN 978-3-0365-1990-6 (Hbk)
ISBN 978-3-0365-1991-3 (PDF)

Contents

About the Editor

Ju-Hyung Kim received B.S. and M.S. degrees in chemical and biological engineering from Seoul National University (Korea) in 2007, and a Ph.D. degree in advanced materials science from the University of Tokyo (Japan) in 2012, working on organic semiconductors and organic/metal interfaces. From 2009 to 2012, he was an IPA Research Fellow of RIKEN (Japan). He also performed postdoctoral work at the Center for Organic Photonics and Electronics Research (OPERA), Kyushu University (Japan). From 2014 to 2016, he worked as an Assistant Professor at Pukyong National University (Korea). Since 2016, he has been working at Ajou University (Korea) as an Associate Professor. His research interests include surface engineering and analysis of organic thin films, organic electronic and optoelectronic applications, and unconventional lithography methodology.

Editorial

Surface and Interface Engineering for Organic Device Applications

Ju-Hyung Kim

Department of Chemical Engineering and Department of Energy Systems Research, Ajou University, Suwon 16499, Korea; juhyungkim@ajou.ac.kr

Citation: Kim, J.-H. Surface and Interface Engineering for Organic Device Applications. *Materials* **2021**, *14*, 4647. https://doi.org/10.3390/ma14164647

Received: 7 August 2021
Accepted: 11 August 2021
Published: 18 August 2021

Publisher's Note: MDPI stays neutral with regard to jurisdictional claims in published maps and institutional affiliations.

In last few decades, organic materials (or carbon-based materials in a broad sense) including polymers have received much attention for their potential applications in electronics, because they have outstanding advantages such as high processibility, mechanical flexibility, and low weight. Extensive research efforts have thus been devoted to the development and advancement of organic materials for various applications, covering a wide range from molecular design to device-fabrication methods. In addition, it has been recognized that surfaces and interfaces play a crucial role in the operation and performance of the devices. For instance, various interactions at organic–metal interfaces are of great importance in organic epitaxy, and also have a strong correlation with intermolecular structures and their electronic properties.

In this context, the main focus of this Special Issue was collecting scientific contributions addressing surface and interface engineering with organic materials, and related applications. The diversity of contributions presented in this Special Issue exhibits the potential of organic materials in a variety of applications that are not limited to the fabrication of organic devices. This Special Issue contains eight featured original research papers as regards nanoarchitecture based on organic elements [1], physical behaviors of organic-based composite materials [2], surface exfoliation and device fabrication using functional polymers [3,4], performance enhancements of organic-based electronic devices [5,6], conductivity analysis of conducting polymers [7], and measurement system applicable to analysis of surface and interface engineering [8].

In the contributed article entitled "Investigation on the Adsorption-Interaction Mechanism of Pb(II) at Surface of Silk Fibroin Protein-Derived Hybrid Nanoflower Adsorbent" [1], Xiang Li et al. demonstrated the preparation of protein-derived hybrid nanoflowers via self-assembly, and investigated their adsorption behaviors and partialized functions of organic and inorganic elements upon adsorption. The prepared nanoflowers exhibited excellent efficiency for Pb(II) removal, which was evaluated by thermodynamic and adsorption kinetics investigation in this study.

Byung Soo Hwang et al. provided an outlook for the development of stimuli-responsive functional materials through their study of gelation behaviors of hydrogels in the contributed article entitled "Thermogelling Behaviors of Aqueous Poly(N-Isopropylacrylamide-co-2-Hydroxyethyl Methacrylate) Microgel–Silica Nano-particle Composite Dispersions" [2]. In this study, an in-depth investigation of the unique thermogelling behaviors of poly(N-isopropylacrylamide)-based microgels through poly(N-isopropylacrylamide-co-2-hydroxyethyl methacrylate) microgel (p(NiPAm-co-HEMA))-silica nanoparticle composite was performed with different molar ratios and various concentrations.

Peng Xiao et al. presented a facile method for the fabrication of liquid metal electrodes using polydimethylsiloxane, and also demonstrated solar-blind photodetection via surface exfoliation in a series of contributed articles entitled "Fabrication of a Flexible Photodetector Based on a Liquid Eutectic Gallium Indium" [3], and "Flexible and Stretchable Liquid Metal Electrodes Working at Sub-Zero Temperature and Their Applications" [4]. Two types

Materials **2021**, *14*, 4647. https://doi.org/10.3390/ma14164647   https://www.mdpi.com/journal/materials

of gallium-based liquid metal, eutectic gallium indium and galinstan, were employed for their works, and comparative studies according to material properties are described in the second article.

In the contributed article entitled "Crack-Assisted Charge Injection into Solvent-Free Liquid Organic Semiconductors via Local Electric Field Enhancement" [5], Ju-Hyung Kim et al. addressed surface engineering of metal electrodes for local electric field enhancements in organic devices. Efficient charge injection into solvent-free liquid organic semiconductors was demonstrated via intentionally cracked metal structures. It was also found that the cracked structures significantly increased the current density, which was strongly supported by a field intensity calculation.

In the contributed article entitled "Cesium Doping for Performance Improvement of Lead(II)-acetate-Based Perovskite Solar Cells" [6], Min-Seok Han et al. presented the preparation of organolead trihalide perovskite materials using lead(II)-acetate as a lead source for which inconvenient antisolvent treatment was not necessary. The effect of cesium doping on the performance of lead(II)-acetate-based solar cells was investigated, and a power conversion efficiency of 18.02% was eventually achieved in this work.

Hyeok Jo Jeong et al. reported the sigmoidal concentration dependence of electrical conductivity of poly(3,4-ethylenedioxythiophene):poly(styrene sulfonate) processed with linear glycol-based additives in the contributed article entitled "Sigmoidal Dependence of Electrical Conductivity of Thin PEDOT:PSS Films on Concentration of Linear Glycols as a Processing Additive" [7]. It was found that repeated ethylene oxides with hydroxyl groups were effective in enhancing the electrical conductivity of poly(3,4-ethylenedioxythiophene):poly(styrene sulfonate) in this work.

In the contributed article entitled "The Methods and Experiments of Shape Measurement for Off-Axis Conic Aspheric Surface" [8], Shijie Li et al. presented three methods (i.e., auto-collimation, single computer-generated hologram, and hybrid compensation) to test the shape accuracy of the off-axis conic aspheric surface with high precision. It was demonstrated that their experimental results involved in peak-to-valley, root-mean-square, and shape distribution were consistent in the three different methods.

In conclusion, the contributed articles in this Special Issue demonstrate relevant progress and the potential of organic materials in a variety of applications. I wish to thank and acknowledge all the authors for their priceless contributions and the editorial members of *Materials*. I am personally grateful to Ms. Freda Zhang, Managing Editor of this Special Issue.

Funding: This research received no external funding.

Data Availability Statement: Not applicable.

Conflicts of Interest: The author declare no conflict of interest.

References

1. Li, X.; Xiong, Y.; Duan, M.; Wan, H.; Li, J.; Zhang, C. Investigation on the Adsorption-Interaction Mechanism of Pb(II) at Surface of Silk Fibroin Protein-Derived Hybrid Nanoflower Adsorbent. *Materials* **2020**, *13*, 1241. [CrossRef] [PubMed]
2. Hwang, B.S.; Kim, J.S.; Kim, J.M.; Shim, T.S. Thermogelling Behaviors of Aqueous Poly(N-Isopropylacrylamide-co-2-Hydroxyethyl Methacrylate) Microgel–Silica Nanoparticle Composite Dispersions. *Materials* **2021**, *14*, 1212. [CrossRef] [PubMed]
3. Xiao, P.; Gwak, H.-J.; Seo, S. Fabrication of a Flexible Photodetector Based on a Liquid Eutectic Gallium Indium. *Materials* **2020**, *13*, 5210. [CrossRef] [PubMed]
4. Xiao, P.; Kim, J.-H.; Seo, S. Flexible and Stretchable Liquid Metal Electrodes Working at Sub-Zero Temperature and Their Applications. *Materials* **2021**, *14*, 4313. [CrossRef] [PubMed]
5. Kim, K.-H.; Park, M.-J.; Kim, J.-H. Crack-Assisted Charge Injection into Solvent-Free Liquid Organic Semiconductors via Local Electric Field Enhancement. *Materials* **2020**, *13*, 3349. [CrossRef] [PubMed]
6. Han, M.-S.; Liu, Z.; Liu, X.; Yoon, J.; Lee, E.-C. Cesium Doping for Performance Improvement of Lead(II)-acetate-Based Perovskite Solar Cells. *Materials* **2021**, *14*, 363. [CrossRef] [PubMed]

7. Jeong, H.J.; Jang, H.; Kim, T.; Earmme, T.; Kim, F.S. Sigmoidal Dependence of Electrical Conductivity of Thin PEDOT:PSS Films on Concentration of Linear Glycols as a Processing Additive. *Materials* **2021**, *14*, 1975. [CrossRef]

8. Li, S.; Zhang, J.; Liu, W.; Liang, H.; Xie, Y.; Li, X. The Methods and Experiments of Shape Measurement for Off-Axis Conic Aspheric Surface. *Materials* **2020**, *13*, 2101. [CrossRef] [PubMed]

materials

Article

Flexible and Stretchable Liquid Metal Electrodes Working at Sub-Zero Temperature and Their Applications

Peng Xiao [1], Ju-Hyung Kim [2,3],* and Soonmin Seo [1],*

[1] Department of Bionano Technology, Gachon University, Seongnam 13120, Korea; zhongpengxiao@gmail.com
[2] Department of Chemical Engineering, Ajou University, Suwon 16499, Korea
[3] Department of Energy Systems Research, Ajou University, Suwon 16499, Korea
* Correspondence: juhyungkim@ajou.ac.kr (J.-H.K.); soonmseo@gachon.ac.kr (S.S.)

Abstract: We investigated characteristics of highly flexible and stretchable electrodes consisting of Galinstan (i.e., a gallium-based liquid metal alloy) under various conditions including sub-zero temperature (i.e., <0 °C) and demonstrated solar-blind photodetection via the spontaneous oxidation of Galinstan. For this work, a simple and rapid method was introduced to fabricate the Galinstan electrodes with precise patterns and to exfoliate their surface oxide layers. Thin conductive films possessing flexibility and stretchability can be easily prepared on flexible substrates with large areas through compression of a dried suspension of Galinstan microdroplets. Furthermore, a laser marking machine was employed to facilitate patterning of the Galinstan films at a high resolution of 20 μm. The patterned Galinstan films were used as flexible and stretchable electrodes. The electrical conductivity of these electrodes was measured to be ~1.3 × 10^6 S m^{-1}, which were still electrically conductive even if the stretching ratio increased up to 130% below 0 °C. In addition, the surface oxide (i.e., Ga_2O_3) layers possessing photo-responsive properties were spontaneously formed on the Galinstan surfaces under ambient conditions, which could be solely exfoliated using elastomeric stamps. By combining Galinstan and its surface oxide layers, solar-blind photodetectors were successfully fabricated on flexible substrates, exhibiting a distinct increase of up to 14.7% in output current under deep ultraviolet irradiation (254 nm wavelength) with an extremely low light intensity of 0.1 mW cm^{-2}, whereas no significant change was observed under visible light irradiation.

Keywords: liquid metals; gallium alloys; Galinstan; flexible electronics photodetectors; solar-blind photodetection

Citation: Xiao, P.; Kim, J.-H.; Seo, S. Flexible and Stretchable Liquid Metal Electrodes Working at Sub-Zero Temperature and Their Applications. *Materials* 2021, 14, 4313. https://doi.org/10.3390/ma14154313

Academic Editor: Ricardo Alcántara

Received: 18 June 2021
Accepted: 30 July 2021
Published: 2 August 2021

Publisher's Note: MDPI stays neutral with regard to jurisdictional claims in published maps and institutional affiliations.

1. Introduction

Recently, liquid metals based on gallium (Ga) alloys have received increasing attention, owing to their outstanding electrical and mechanical properties [1–3]. The Ga-based metal alloys that exist as virtually non-toxic liquids at room temperature show not only excellent stretchability and deformability but also environmental friendliness and recyclability. In this context, extensive research efforts have been devoted to the development of various applications using Ga-based metal alloys, such as sensors [4], reconfigurable antennas [5], and soft electrodes [6]. These fluidic metal alloys also show great potential for electronic skins [7,8] and wearable electronics [9–11]. Among various Ga-based metal alloys, Galinstan (68.5% Ga, 21.5% indium (In), and 10% tin (Sn)) has been notably studied in recent years due to its remarkably low toxicity and melting point (~−19 °C) [12], which is also suitable for flexible and stretchable devices operating below 0 °C compared to other eutectic gallium indium (EGaIn).

Although Galinstan shows outstanding properties including flexibility and stretchability even under cold conditions, its high surface tension and rapid oxidation rate hinder the fabrication of desirable patterns for electronic devices and circuits in comparison with other functional materials [13–15]. Various methods for Galinstan patterning thus have been developed and enhanced, including microfluidic injection [16–21], photolithography [22,23],

Materials 2021, 14, 4313. https://doi.org/10.3390/ma14154313

stencil lithography [24–27], imprint lithography [28], microcontact printing [29,30], and composite material synthesis [31]. Each of these methods has individual advantages (i.e., high processability, high resolution limits, high stability, or cost-effective fabrication); however, integrating all the advantageous elements is still a challenge. For instance, the demanding process conditions for delaying surface oxidation or preventing leakage of the liquid metal or the low patterning resolution limits still need to be improved depending on the method. Laser ablation is one of the patterning methods with high processability, enabling the rapid fabrication of electrodes with complex features [32,33]. Although the pattern resolution is limited by the beam spot of laser, which is normally in the range of tens to hundreds of micrometers, this method can be directly employed for various applications without laborious pre- and/or post-treatments.

In addition, it is worth noting that the spontaneous oxidation of Galinstan in air leads to the formation of thin Ga oxide (Ga_2O_3) films on Galinstan surfaces [3,34,35]. Ga_2O_3 with a wide bandgap (~4.9 eV), which is rapidly formed on Galinstan surfaces in less than one second during the patterning processes in air [2], is transparent in the visible light region and exhibits high light-absorption coefficients in the deep ultraviolet (UV) region [35–38]. The surface oxide layers normally degrade the metallic properties of Galinstan; however, these layers are also expected to be utilized for solar-blind photodetection (i.e., deep UV detection insensitive to solar radiation) if they can be neatly separated from the bulk material [39].

Herein, we investigated characteristics of the Galinstan electrodes to verify flexibility and stretchability under various conditions including sub-zero temperature (i.e., <0 °C). For this study, a simple and rapid method was employed to fabricate the Galinstan electrodes with precise patterns. Thin Galinstan films with high electrical conductivity were uniformly deposited on flexible polydimethylsiloxane (PDMS) substrates by the compression of Galinstan microdroplets and sequentially patterned using a fiber laser marking machine. The transparent PDMS substrates were found to be undamaged by a laser with a wavelength of 1064 nm, and only the Galinstan layers were ablated according to the designed electrode shapes. In addition, the surface oxide (i.e., Ga_2O_3) layers of the Galinstan electrodes were also examined to confirm their potential for solar-blind photodetection. For the photoactive components, the thin Ga_2O_3 films, spontaneously formed on the Galinstan surfaces, were exfoliated using elastomeric PDMS stamps [39,40] and then transferred onto the patterned Galinstan electrodes to complete the device structure for solar-blind photodetection. By combining Galinstan and Ga_2O_3 films, sensitive solar-blind photodetectors were successfully fabricated on flexible substrates. The photodetectors showed a distinct increase of up to ~15.1% in output current under deep UV irradiation (254 nm wavelength) with an extremely low light intensity of 0.1 mW cm^{-2}, whereas no significant change was observed under visible light irradiation.

2. Materials and Methods

2.1. Preparation of Galinstan Microdroplets

Galinstan (68.5 wt.% Ga, 21.5 wt.% In, and 10.0 wt.% Sn) and PDMS (SYLGARD 184) were purchased from Geratherm Medical AG (Geratal, Germany) and DOW (Midland, MI, USA), respectively. EGaIn (75.5 wt.% Ga and 24.5 wt.% In) was purchased from Sigma-Aldrich Korea (Seoul, Korea). To prepare the Galinstan microdroplets, 0.5 g of Galinstan was sonicated in ethanol for 30 min (80 W, 40 KHz). Galinstan was well dispersed during the sonication process and rapidly stabilized by surface oxidation in ethanol, resulting in the suspension of Galinstan microdroplets (<10 μm) as shown in Figure 1a. The same procedure was repeatedly performed for the preparation of the EGaIn microdroplets.

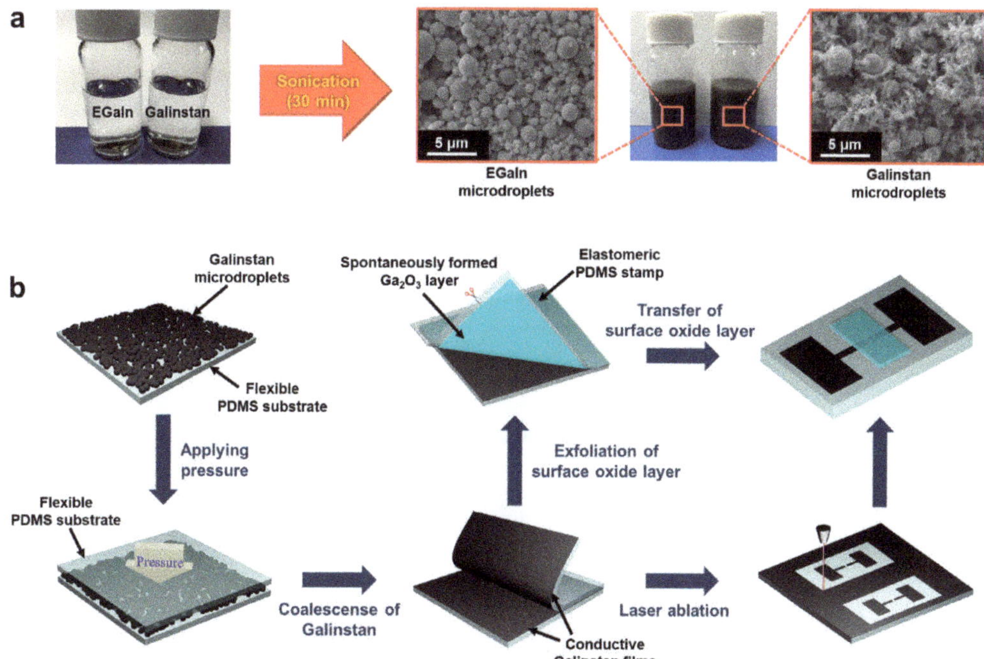

Figure 1. (**a**) Preparation of suspensions comprising EGaIn and Galinstan microdroplets by sonication. (**b**) Schematic illustration of the fabrication of flexible solar-blind photodetector using Galinstan microdroplets.

2.2. Preparation of Elastomeric PDMS Substrates and Stamps

For the PDMS substrates, the PDMS precursor, consisting of a silicone elastomer base and a curing agent (in a 10:1 weight ratio), was poured onto a flat Petri dish, and subsequently degassed in a vacuum desiccator for 1 h. The sample was cured at 80 °C for 1 h in a convection oven. After thermal curing, the PDMS film was easily peeled off from the Petri dish and then cut into 50 mm × 50 mm specimens. The thickness of each substrate was measured as ~1 mm. Sticky elastomeric PDMS stamps were individually prepared to exfoliate the thin Ga_2O_3 films. The mixing ratio of the PDMS precursor was modified to a 11:1 weight ratio to delay the saturation of cross-linking and enhance its adhesive properties, and the same preparation procedure as for the PDMS substrates was followed.

2.3. Fabrication of Patterned Galinstan Electrodes

The suspension containing the Galinstan microdroplets was drop-dispensed onto the flat PDMS substrate and then slowly dried at 30 °C for 24 h to avoid the formation of structural defects induced by rapid evaporation of the solvent. The dried suspension of the microdroplets in thin-film form was covered with another flat PDMS substrate and subsequently pressed at a pressure of 15 MPa for 5 s to collapse the surface oxide layers and connect Galinstan. After releasing the pressure, the upper PDMS mold was peeled off from the bottom PDMS substrate, resulting in the formation of thin conductive Galinstan films deposited on both PDMS substrates.

The Galinstan films were patterned using a laser marking machine (50 W, Dongil Laser Technology, Gwangju, Korea). The scanning speed of the laser marking machine was 600 mm s^{-1}, and the power intensity was 1.0 % of its maximum power (i.e., 0.5 W). A high resolution of 20 µm was achieved in the Galinstan patterning process by this laser ablation method.

2.4. Characterization of Flexible and Stretchable Liquid Metal Electrodes

To investigate characteristics of the liquid metal electrodes under various conditions, thin conductive films (15×25 mm^2) of Galinstan and EGaIn were individually prepared with the same procedure. The thickness of each film was 1 µm. A semiconductor characterization system (4200-SCS, Keithley, Beaverton, OR, USA) was used for the measurements.

2.5. Fabrication of Flexible Solar-Blind Photodetectors Using Ga_2O_3

The elastomeric PDMS stamp was brought into contact with the surface of the Galinstan film. A thin Ga_2O_3 film (<10 nm), which was spontaneously formed on the Galinstan surface, was attached to the sticky PDMS stamp and easily exfoliated from the Galinstan film by peeling off the stamp. The transparent Ga_2O_3 film on the PDMS stamp was cut and placed between two patterned Galinstan electrodes to complete the device structure.

2.6. Measurements

A semiconductor parameter analyzer (Keithley 4200, Beaverton, OR, USA) and resistivity meter (Loresta-GX MCP-T700, Mitsubishi Chemical Analytech, Yamato, Japan) were used to measure electrical properties and perform bending tests on the devices. Photocurrent measurements were performed for deep UV and visible regions using a UV lamp (8 W, Vilber Lourmat, Collégien, France) and a halogen lamp (FOK-100W, Fiber Optic Korea, Cheonan, Korea). The surface morphology was also investigated using atomic force microscopy (AFM; Nanoscope IIIa, Digital Instruments, Bresso, Italy) and scanning electron microscopy (SEM; JSM-7500F, Tokyo, Japan).

3. Results and Discussion

Suspensions comprising microdroplets of Galinstan and EGaIn were individually prepared as shown in Figure 1a. It exhibited a matt dark gray color due to diffuse reflections and surface oxide layers of the microdroplets. As schematically illustrated in Figure 1b, the suspension was drop-dispensed onto a flat PDMS substrate and then slowly dried at 30 °C for 24 h. As the slow drying process hindered the formation of structural defects induced by rapid evaporation of the solvent, the microdroplets were densely aggregated in thin-film form with high uniformity. For Galinstan, the size of each microdroplet was less than 5 µm, and rod-shaped particles were also observed between the rounded droplets. It is well known that rounded Galinstan microdroplets are surrounded by a thin layer of carbon and Ga_2O_3, of which the inner core is composed of Ga, In, and Sn [41,42]. As previously reported, the rod-shaped particles possibly consisted of Ga oxide monohydroxide ((GaO)OH) [43]. Note that Ga in Ga-based metal alloys can react with decomposed OH^- in the presence of O_2, leading to the crystallization of (GaO)OH as follows:

$$2Ga + 2OH^- + O_2 \rightarrow 2(GaO)OH \tag{1}$$

The amount of rod-shaped particles is significantly less than that of round particles, and it is expected to be further reduced at low-temperature conditions because the crystallization strongly depends on heat and reactive oxygen species originating from sonication [44,45].

The thin film comprising aggregated Galinstan microdroplets was not electrically conductive because each droplet was fully covered by a non-conductive Ga_2O_3 layer. Thus, another flat PDMS substrate for protecting Galinstan was brought into contact with the thin film, and an external pressure of 15 MPa was sequentially applied to the sample to collapse the surface Ga_2O_3 layers and connect Galinstan. With the collapse of the surface Ga_2O_3 layers, the Galinstan microdroplets were connected to achieve a continuous phase between the two PDMS substrates. After peeling off the upper PDMS substrate from the bottom PDMS substrate, thin Galinstan films were consequently formed on both PDMS substrates (see Figure 2a). The thickness of each glossy film was measured to be less than 1 µm, of which the surface partially cracked due to rapid surface oxidation during the peeling process.

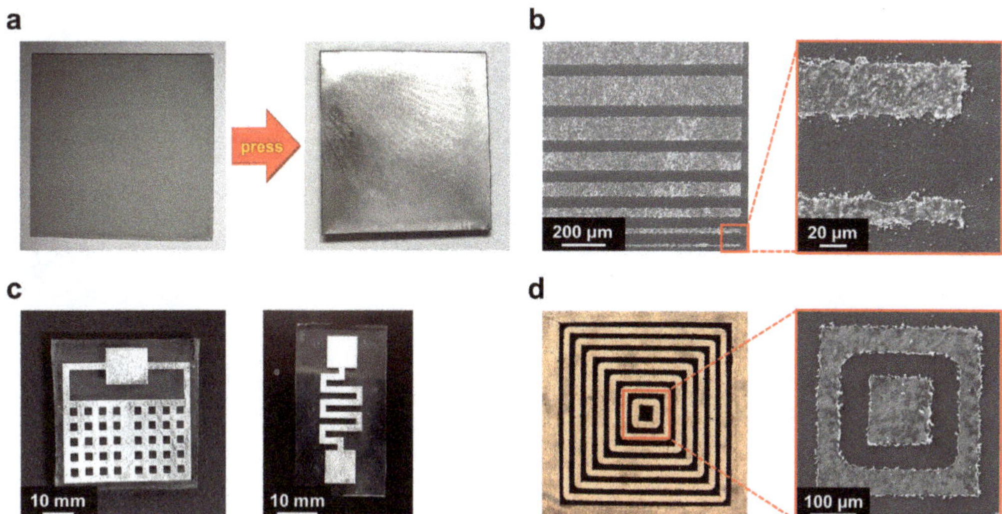

Figure 2. (**a**) Photographs of (left) a dried suspension of Galinstan microdroplets in thin-film form on a flexible PDMS substrate, and (right) a conductive Galinstan film fabricated by compression and separation using PDMS. (**b**) SEM images of patterned Galinstan films, of which the minimum line width is ~20 µm. (**c**) Photographs of exemplary Galinstan films, patterned by laser ablation. (**d**) Optical microscopy image and its close-up SEM image of an exemplary Galinstan structure with complex pattern.

A fiber laser marking machine (λ ~ 1064 nm) was employed for the direct patterning of the Galinstan films, enabling the fabrication of accurate and desirable features with sub-100 µm resolution. Various exemplary features of the patterned Galinstan films are shown in Figure 2b–d. The smallest feature size of the Galinstan line was 20 µm. Laser ablation is a fast and precise method for patterning liquid metal electrodes, facilitating the fabrication of complex and hollow patterns. In addition, this light-based patterning method does not cause significant damage to the transparent substrates, such as glass and PDMS, which do not directly absorb the energy from a fiber laser. After completing the laser ablation process, partial buckling was observed on the PDMS substrates.

The electrical resistance and conductivity of the Galinstan film were measured corresponding to the structural deformation of the PDMS substrate. For the measurements of electrical properties, the patterned Galinstan electrode (80 µm × 5 mm) was used as shown in Figure 3. Its thickness was less than 1 µm. The electrical resistance and conductivity were initially measured as 48.3 Ω and ~1.3×10^6 S m^{-1}, respectively. In response to the deformation ratio, the electrical resistance gradually increased to 73.8 Ω, corresponding to an electrical conductivity of ~8.5×10^5 S m^{-1}. In comparison with pure Galinstan, in which the electrical conductivity was found to be ~3.5×10^6 S m^{-1}, the relatively low electrical conductivity of the Galinstan films used in this work could be attributed to the partial cracks and insulating components remaining in the films, such as (GaO)OH and Ga$_2$O$_3$. However, since the majority of the film components were Galinstan, the fabricated films still possessed electrical conductivity high enough to be used as flexible electrodes. It should be noted that the electrical conductivity of the fabricated Galinstan film (~1.3×10^6 S m^{-1}) is slightly lower than that of the thin EGaIn film (~2.2×10^6 S m^{-1}) prepared using the same procedure [39]. It is possibly originating from the content of the insulating material in the suspension. As shown in Figure 1a, the content of the rod-shaped particles in the Galinstan suspension is significantly higher than in the EGaIn suspension under our experimental conditions, which may cause a decrease in overall electrical conductivity.

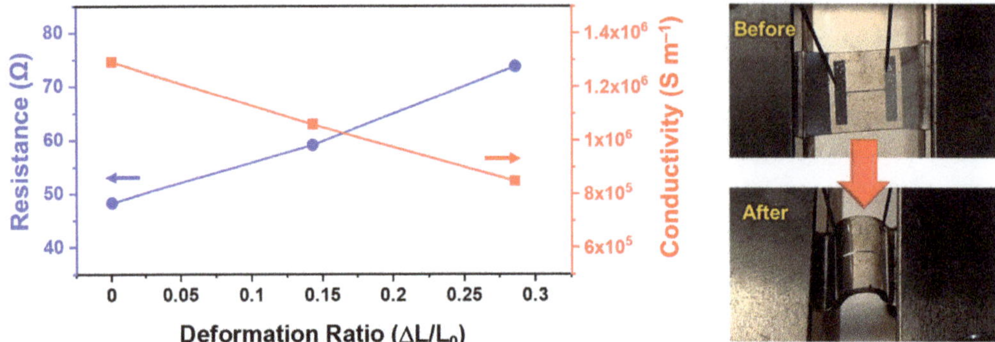

Figure 3. Electrical resistance and conductivity of a line-patterned Galinstan electrode (80 μm × 5 mm) corresponding to the structural deformation of the PDMS substrate.

One advantageous property of Galinstan is the liquid phase, maintaining its flexibility and stretchability, even below 0 °C. To compare with EGaIn, electrical resistances of the two materials were measured corresponding to lateral stretching (up to ~130%) of the PDMS substrates. For the measurements, the flat Galinstan and EGaIn electrodes were individually prepared on the PDMS substrates ($15 \times 25 \times 1$ mm^3) and then stretched up to 130% at room temperature and -10 °C, respectively. Changes in the electrical resistances upon lateral stretching are shown in Figure 4a. At room temperature, the electrical resistances of both materials slightly increased with the stretching ratios, which is possibly originating from structural deformation [26]. At the temperature of -10 °C, the cracks were generated inside the EGaIn film upon the lateral stretches (see Figure 4b), leading to significant reduction in the film continuity. When the stretching ratio was above 110%, the electrical conductivity of the EGaIn film was thus not observed. However, differently from EGaIn, the Galinstan film was still electrically conductive even if the stretching ratio increased up to 130%. These results were caused by the difference in the melting points of the two materials (i.e., ~ -19 °C for Galinstan and ~16 °C for EGaIn). At -10 °C, the EGaIn film in the solid phase was significantly damaged, whereas the Galinstan film in the liquid phase showed excellent film continuity (see Figure 4c).

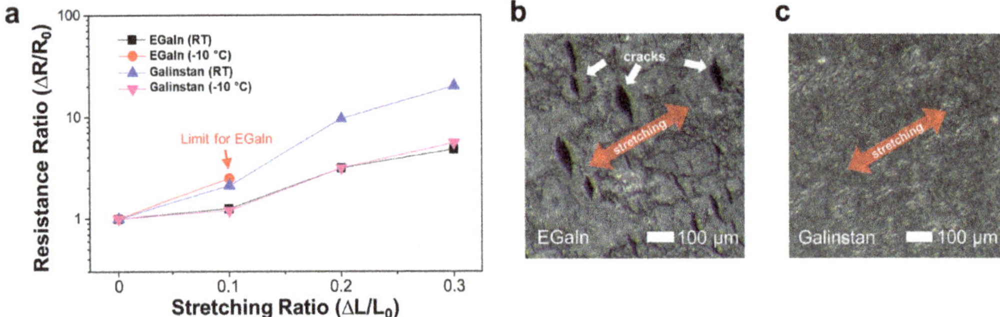

Figure 4. (**a**) Changes in electrical resistances of EGaIn and Galinstan films corresponding to lateral stretching at varied temperatures (i.e., room temperature (RT) and -10 °C). (**b**) SEM image of a cracked EGaIn film after lateral stretching below 0 °C. (**c**) SEM image of a stretched Galinstan film below 0 °C, exhibiting excellent film continuity.

In addition, the Ga$_2$O$_3$ layer, which was spontaneously formed on the Galinstan film, was neatly exfoliated using an elastomeric PDMS stamp for further investigation. It should be noted that inherently high adhesion between the thin oxide shell and PDMS was reported [42], and the elastomeric PDMS stamps enabled intimate contact with the oxide surfaces. In this work, the mixing ratio of the PDMS precursor was further modified to enhance the adhesive properties, demonstrating excellent contact characteristics with a relatively rough Ga_2O_3 surface. The elastomeric PDMS stamp was placed on the surface of the Galinstan film without applying any external pressure and was then detached. In this process, the transparent Ga_2O_3 layer was successfully transferred onto the stamp. The transferred Ga_2O_3 film was slightly darker than the bare PDMS substrate because small Galinstan residues in the form of islands (<10 μm) remained on the substrate, as shown in Figure 5c. However, all Galinstan residues were entirely isolated from each other and wrapped with Ga_2O_3, resulting in the formation of a non-metallic film. The thickness of the exfoliated Ga_2O_3 film was measured as ~13 nm using AFM, as shown in Figure 5b. It is worth noting that the measured thickness of the Ga_2O_3 film in this work is thicker than that of the single surface oxide layer of Galinstan (i.e., ~3 nm) due to further oxidation during the exfoliation process. The surface roughness and embossed features of the fabricated Galinstan film, as shown in Figure 2d, could also affect the thickness of the Ga_2O_3 film.

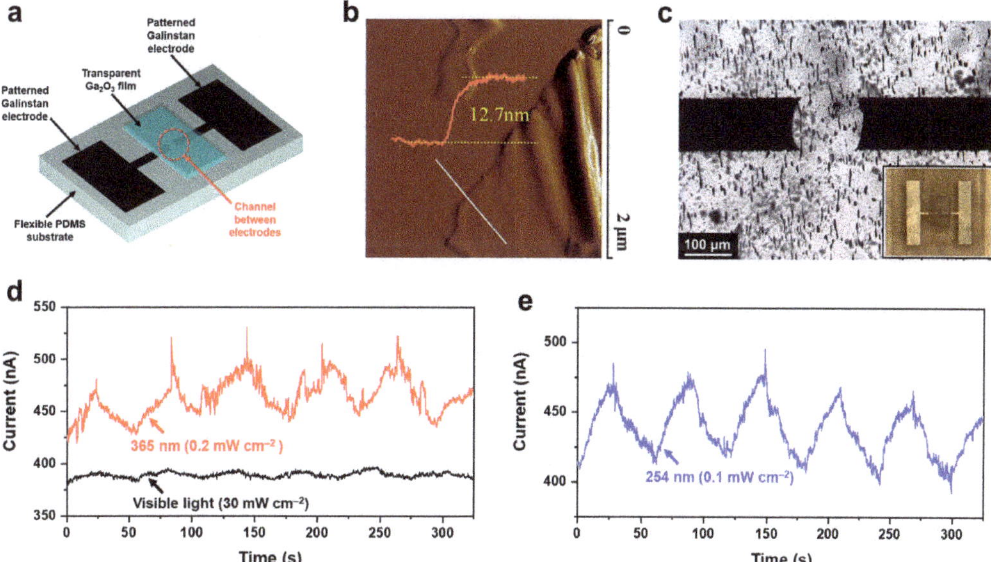

Figure 5. (**a**) Schematic illustration of solar-blind photodetector. (**b**) AFM image of transparent Ga_2O_3 film, exfoliated using a PDMS stamp. The height profile along the white line is also indicated. (**c**) Optical microscopy image of a channel between two patterned Galinstan electrodes. The black areas correspond to Galinstan beneath the transparent Ga_2O_3 film. Photograph of the fabricated photodetector is also shown in the inset. (**d**) Output characteristics of the solar-blind photodetector under irradiations of visible light (with a halogen lamp; ranging from 350 to 900 nm) and UV light of 365 nm wavelength, respectively. (**e**) Output characteristics of the solar-blind photodetector under deep UV irradiation (254 nm wavelength). The on/off switching of each irradiation was manually performed at 30 s intervals, and the output characteristics were constantly measured at a sample bias voltage of 0.1 V.

To investigate the solar-blind photodetective properties of the exfoliated Ga_2O_3 film in consideration of its wide bandgap (~4.9 eV) [35–38], a channel between two patterned Galinstan electrodes was bridged using Ga_2O_3, as shown in Figure 5a. For this work, the conductive Galinstan film, prepared on a large area (50 mm × 50 mm), was patterned by laser ablation to form a channel. The transparent Ga_2O_3 film, individually prepared on the PDMS stamp, was placed between the channel to complete the device structure (see Figure 5b). The output current was constantly measured at a sample bias voltage of 0.1 V. Under visible light irradiation with a halogen lamp (ranging from 350 to 900 nm), the output current only increased by ~2.7% at a high light intensity of 30 mW cm^{-2} (see

Figure 5c), and no significant change in the output current was observed at lower light intensities. As the contribution of the short-wavelength region in the emission spectrum is not negligible at high intensity, a small increase in the output current could be detected at light intensities above 30 mW cm^{-2}. To confirm this speculation, the output current was also measured under irradiation of UV light of a 365 nm wavelength, which was contained in the emission spectrum of the halogen lamp, and the output current in effect increased by ~14.9% at a low light intensity of 0.2 mW cm^{-2}. Eventually, under deep UV irradiation (254 nm wavelength) with an extremely low light intensity of 0.1 mW cm^{-2}, the output current sensitively increased by up to 15.1% (see Figure 5d). These results strongly suggest that the combination of Galinstan and its surface oxide layers can be used for sensitive solar-blind photodetectors that possess remarkable advantages, such as low-cost and easy processability under ambient conditions, and flexibility.

4. Conclusions

We investigated characteristics of the flexible and stretchable Galinstan electrodes under various conditions including sub-zero temperature (i.e., <0 °C) and successfully demonstrated solar-blind photodetection via the spontaneous oxidation of Galinstan. In this work, a simple and rapid method was introduced for fabricating the flexible and stretchable Galinstan electrodes with precise patterns and exfoliating the surface oxide layers to complete the device structure enabling solar-blind photodetection. A suspension consisting of Galinstan microdroplets was prepared by sonication. Thin Galinstan films with thickness less than 1 μm were uniformly deposited on flexible PDMS substrates by compression of the dried suspension of the microdroplets. The Galinstan films, deposited on a large area (50 mm × 50 mm), were sequentially patterned using a fiber laser marking machine (λ~1064 nm), and accurate and desirable features with a high resolution of 20 μm were fabricated. Although the electrical conductivity of the fabricated films was lower than that of pure Galinstan, they still possessed electrical conductivity high enough to be used as flexible and stretchable electrodes even below 0 °C. For the photoactive components, thin Ga_2O_3 layers, spontaneously formed on the Galinstan surfaces, were exfoliated using elastomeric PDMS stamps and successfully transferred onto the patterned Galinstan electrodes to complete the device structure for solar-blind photodetection. The solar-blind photodetectors demonstrated a distinct increase of up to ~15.1% in the output current under deep UV irradiation (254 nm wavelength) with an extremely low light intensity of 0.1 mW cm^{-2}, whereas no significant change was observed under visible light irradiation. These results strongly suggest that Galinstan can be used for flexible and stretchable electrodes working under extreme conditions, and the combination with its surface oxide layer also shows great potential for sensitive solar-blind photodetectors that possess outstanding advantages, such as low-cost and easy processability under ambient conditions. We anticipate that these results will contribute to the development of flexible and stretchable electronic devices based on liquid metals, which can lead to further application of sensors under extreme conditions.

Author Contributions: P.X.: conceptualization, methodology, formal analysis, investigation, data curation, visualization, and writing—original draft preparation; J.-H.K.: investigation, data curation, visualization, writing—original draft preparation, review and editing, and supervision; S.S.: conceptualization, formal analysis, visualization, writing—review and editing, supervision, and funding acquisition. All authors have read and agreed to the published version of the manuscript.

Funding: This work was supported by the Basic Science Research Program through the National Research Foundation of Korea (NRF-2021R1F1A1047036 and NRF-2018R1D1A1B07041253).

Institutional Review Board Statement: Not applicable.

Informed Consent Statement: Not applicable.

Data Availability Statement: Data sharing is not applicable to this article.

Conflicts of Interest: The authors declare no conflict of interest.

References

1. Zavabeti, A.; Ou, J.Z.; Carey, B.J.; Syed, N.; Orrell-Trigg, R.; Mayes, E.L.H.; Xu, C.; Kavehei, O.; O'Mullane, A.P.; Kaner, R.B.; et al. A liquid metal reaction environment for the room-temperature synthesis of atomically thin metal oxides. *Science* **2017**, *358*, 332–335. [CrossRef] [PubMed]
2. Daeneke, T.; Khoshmanesh, K.; Mahmood, N.; de Castro, I.A.; Esrafilzadeh, D.; Barrow, S.J.; Dickey, M.; Kalantar-Zadeh, K. Liquid metals: Fundamentals and applications in chemistry. *Chem. Soc. Rev.* **2018**, *47*, 4073–4111. [CrossRef]
3. Carey, B.J.; Ou, J.Z.; Clark, R.M.; Berean, K.J.; Zavabeti, A.; Chesman, A.S.; Russo, S.P.; Lau, D.W.; Xu, Z.-Q.; Bao, Q.; et al. Wafer-scale two-dimensional semi-conductors from printed oxide skin of liquid metals. *Nat. Commun.* **2017**, *8*, 1–10.
4. Kim, M.; Seo, S. Flexible pressure and touch sensor with liquid metal droplet based on gallium alloys. *Mol. Cryst. Liq. Cryst.* **2019**, *685*, 40–46. [CrossRef]
5. Elassy, K.S.; Akau, T.K.; Shiroma, W.A.; Seo, S.; Ohta, A.T. Low-cost rapid fabrication of conformal liquid-metal patterns. *Appl. Sci.* **2019**, *9*, 1565. [CrossRef]
6. Yu, Y.; Wang, Q.; Wu, Y.H.; Liu, J. Liquid metal soft electrode triggered discharge plasma in aqueous solution. *RSC Adv.* **2016**, *6*, 114773–114778. [CrossRef]
7. Hong, Y.J.; Jeong, H.; Cho, K.W.; Lu, N.; Kim, D. Wearable and implantable devices for cardiovascular healthcare: From monitoring to therapy based on flexible and stretchable electronics. *Adv. Funct. Mater.* **2019**, *29*. [CrossRef]
8. Chen, L.Y.; Tee, B.C.-K.; Chortos, A.L.; Schwartz, G.; Tse, V.; Lipomi, D.J.; Wong, H.-S.P.; McConnell, M.V.; Bao, Z. Continuous wireless pressure monitoring and mapping with ultra-small passive sensors for health monitoring and critical care. *Nat. Commun.* **2014**, *5*, 5028. [CrossRef] [PubMed]
9. Wang, S.; Xu, J.; Wang, W.; Wang, G.-J.N.; Rastak, R.; Molina-Lopez, F.; Chung, J.W.; Niu, S.; Feig, V.R.; Lopez, J.; et al. Skin electronics from scalable fabrication of an intrinsically stretchable transistor array. *Nat. Cell Biol.* **2018**, *555*, 83–88. [CrossRef] [PubMed]
10. Yang, J.C.; Mun, J.; Kwon, S.Y.; Park, S.; Bao, Z.; Park, S. Electronic skin: Recent progress and future prospects for skin-attachable devices for health monitoring, robotics, and prosthetics. *Adv. Mater.* **2019**, *31*, 1904765. [CrossRef]
11. Yang, Y.; Han, J.; Huang, J.; Sun, J.; Wang, Z.L.; Seo, S.; Sun, Q. Stretchable energy-harvesting tactile interactive interface with liquid-metal-nanoparticle-based electrodes. *Adv. Funct. Mater.* **2020**, *30*, 1909652. [CrossRef]
12. Liu, T.; Sen, P.; Kim, C.-J. Characterization of nontoxic liquid-metal alloy galinstan for applications in microdevices. *J. Microelectromechanical Syst.* **2011**, *21*, 443–450. [CrossRef]
13. Shin, J.; Jeong, B.; Kim, J.; Nam, V.B.; Yoon, Y.; Jung, J.; Hong, S.; Lee, H.; Eom, H.; Yeo, J.; et al. Sensitive wearable temperature sensor with seamless monolithic integration. *Adv. Mater.* **2020**, *32*, 1905527. [CrossRef] [PubMed]
14. Kim, J.Y.; Oh, J.Y.; Lee, T.I. Multi-dimensional nanocomposites for stretchable thermoelectric applications. *Appl. Phys. Lett.* **2019**, *114*, 043902. [CrossRef]
15. Suh, Y.D.; Jung, J.; Lee, H.; Yeo, J.; Hong, S.; Lee, P.; Lee, D.; Ko, S.H. Nanowire reinforced nanoparticle nanocomposite for highly flexible transparent electrodes: Borrowing ideas from macrocomposites in steel-wire reinforced concrete. *J. Mater. Chem. C* **2016**, *5*, 791–798. [CrossRef]
16. So, J.-H.; Dickey, M. Inherently aligned microfluidic electrodes composed of liquid metal. *Lab Chip* **2011**, *11*, 905–911. [CrossRef]
17. Park, Y.-L.; Chen, B.-R.; Wood, R.J. Design and fabrication of soft artificial skin using embedded microchannels and liquid conductors. *IEEE Sens. J.* **2012**, *12*, 2711–2718. [CrossRef]
18. So, J.-H.; Koo, H.-J.; Dickey, M.D.; Velev, O.D. Ionic current rectification in soft-matter diodes with liquid-metal electrodes. *Adv. Funct. Mater.* **2011**, *22*, 625–631. [CrossRef]
19. Parekh, D.; Ladd, C.; Panich, L.; Moussa, K.; Dickey, M.D. 3D printing of liquid metals as fugitive inks for fabrication of 3D microfluidic channels. *Lab Chip* **2016**, *16*, 1812–1820. [CrossRef]
20. Yang, Y.; Sun, N.; Wen, Z.; Cheng, P.; Zheng, H.; Shao, H.; Xia, Y.; Chen, C.; Lan, H.; Xie, X.; et al. Liquid-metal-based super-stretchable and structure-designable triboelectric nanogenerator for wearable electronics. *ACS Nano* **2018**, *12*, 2027–2034. [CrossRef] [PubMed]
21. Zhang, R.; Ye, Z.; Gao, M.; Gao, C.; Zhang, X.; Li, L.; Gui, L. Liquid metal electrode-enabled flexible microdroplet sensor. *Lab Chip* **2020**, *20*, 496–504. [CrossRef]
22. Park, C.W.; Moon, Y.G.; Seong, H.; Jung, S.W.; Oh, J.-Y.; Na, B.S.; Park, N.-M.; Lee, S.S.; Im, S.G.; Koo, J.B. Photolithog-raphy-based patterning of liquid metal interconnects for monolithically integrated stretchable circuits. *ACS Appl. Mater. Interfaces* **2016**, *8*, 15459–15465. [CrossRef] [PubMed]
23. Moon, Y.G.; Koo, J.B.; Park, N.-M.; Oh, J.-Y.; Na, B.S.; Lee, S.S.; Ahn, S.-D.; Park, C.W. Freely deformable liquid metal grids as stretchable and transparent electrodes. *IEEE Trans. Electron Devices* **2017**, *64*, 5157–5162. [CrossRef]
24. Wang, L.; Liu, J. Pressured liquid metal screen printing for rapid manufacture of high resolution electronic patterns. *RSC Adv.* **2015**, *5*, 57686–57691. [CrossRef]
25. Lazarus, N.; Bedair, S.S.; Kierzewski, I.M. Ultrafine pitch stencil printing of liquid metal alloys. *ACS Appl. Mater. Interfaces* **2017**, *9*, 1178–1182. [CrossRef]
26. Park, T.H.; Kim, J.; Seo, S. Facile and rapid method for fabricating liquid metal electrodes with highly precise patterns via one-step coating. *Adv. Funct. Mater.* **2020**, *30*. [CrossRef]

27. Dong, R.; Wang, L.; Hang, C.; Chen, Z.; Liu, X.; Zhong, L.; Qi, J.; Huang, Y.; Liu, S.; Wang, L.; et al. Printed stretchable liquid metal electrode arrays for in vivo neural recording. *Small* **2021**, *17*, 2006612. [CrossRef]
28. Kim, M.-G.; Brown, D.K.; Brand, O. Nanofabrication for all-soft and high-density electronic devices based on liquid metal. *Nat. Commun.* **2020**, *11*, 1–11. [CrossRef] [PubMed]
29. Tabatabai, A.; Fassler, A.; Usiak, C.; Majidi, C. Liquid-phase gallium–indium alloy electronics with microcontact printing. *Langmuir* **2013**, *29*, 6194–6200. [CrossRef]
30. Kim, J.H.; Seo, S. Fabrication of an imperceptible liquid metal electrode for triboelectric nanogenerator based on gallium alloys by contact printing. *Appl. Surf. Sci.* **2020**, *509*, 145353. [CrossRef]
31. Ma, Z.; Huang, Q.; Xu, Q.; Zhuang, Q.; Zhao, X.; Yang, Y.; Qiu, H.; Yang, Z.; Wang, C.; Chai, Y.; et al. Permeable superelastic liquid-metal fibre mat enables biocompatible and monolithic stretchable electronics. *Nat. Mater.* **2021**, *20*, 859–868. [CrossRef] [PubMed]
32. Nam, V.B.; Giang, T.T.; Koo, S.; Rho, J.; Lee, D. Laser digital patterning of conductive electrodes using metal oxide nano-materials. *Nano Converg.* **2020**, *7*, 23. [CrossRef]
33. Nam, V.B.; Shin, J.; Yoon, Y.; Giang, T.T.; Kwon, J.; Suh, Y.D.; Yeo, J.; Hong, S.; Ko, S.H.; Lee, D. Highly stable Ni-based flexible transparent conducting panels fabricated by laser digital patterning. *Adv. Funct. Mater.* **2019**, *29*. [CrossRef]
34. Kim, D.; Thissen, P.; Viner, G.; Lee, D.-W.; Choi, W.; Chabal, Y.J.; Lee, J.-B. Recovery of nonwetting characteristics by surface modification of gallium-based liquid metal droplets using hydrochloric acid vapor. *ACS Appl. Mater. Interfaces* **2012**, *5*, 179–185. [CrossRef]
35. Akbari, M.K.; Hai, Z.; Wei, Z.; Ramachandran, R.K.; Detavernier, C.; Patel, M.; Kim, J.; Verpoort, F.; Lu, H.; Zhuiykov, S. Sonochemical functionalization of the low-dimensional surface oxide of Galinstan for heterostructured optoelectronic applications. *J. Mater. Chem. C* **2019**, *7*, 5584–5595. [CrossRef]
36. Li, Y.; Tokizonno, T.; Liao, M.; Zhong, M.; Koide, Y.; Yamada, I.; Delaunay, J.-J. Efficient assembly of bridged β-Ga2O3 nan-owires for solar-blind photodetection. *Adv. Funct. Mater.* **2010**, *20*, 3972–3978. [CrossRef]
37. Hu, G.C.; Shan, C.X.; Zhang, N.; Jiang, M.M.; Wang, S.-P.; Shen, D.Z. High gain Ga_2O_3 solar-blind photodetectors realized via a carrier multiplication process. *Opt. Express* **2015**, *23*, 13554–13561. [CrossRef] [PubMed]
38. Kwon, Y.; Lee, G.; Oh, S.; Kim, J.; Pearton, S.J.; Ren, F. Tuning the thickness of exfoliated quasi-two-dimensional β-Ga2O3 flakes by plasma etching. *Appl. Phys. Lett.* **2017**, *110*, 131901. [CrossRef]
39. Xiao, P.; Gwak, H.-J.; Seo, S. Fabrication of a flexible photodetector based on a liquid eutectic gallium indium. *Materials* **2020**, *13*, 5210. [CrossRef]
40. Hwang, W.S.; Verma, A.; Peelaers, H.; Protasenko, V.; Rouvimov, S.; Xing, H.; Seabaugh, A.; Haensch, W.; Walle, C.V.; Galazka, Z.; et al. High-voltage field effect transistors with wide-bandgap β-Ga2O3 nanomembranes. *Appl. Phys. Lett.* **2014**, *104*, 203111. [CrossRef]
41. Boley, J.W.; White, E.L.; Kramer, R.K. Mechanically sintered gallium-indium nanoparticles. *Adv. Mater.* **2015**, *27*, 2355–2360. [CrossRef] [PubMed]
42. Lin, Y.; Cooper, C.; Wang, M.; Adams, J.J.; Genzer, J.; Dickey, M.D. Handwritten, soft circuit boards and antennas using liquid metal nanoparticles. *Small* **2015**, *11*, 6397–6403. [CrossRef] [PubMed]
43. Mingkuan, Z.; Siyuan, Y.; Wei, R.; Jing, L. Transformable soft liquid metal micro/nanomaterials. *Mater. Sci. Eng. R Rep.* **2019**, *138*, 1–35.
44. Lin, Y.; Liu, Y.; Genzer, J.; Dickey, M.D. Shape-transformable liquid metal nanoparticles in aqueous solution. *Chem. Sci.* **2017**, *8*, 3832–3837. [CrossRef] [PubMed]
45. Lu, Y.; Lin, Y.; Chen, Z.; Hu, Q.; Liu, Y.; Yu, S.; Gao, W.; Dickey, M.D.; Gu, Z. Enhanced endosomal escape by light-fueled liquid-metal transformer. *Nano Lett.* **2017**, *17*, 2138–2145. [CrossRef] [PubMed]

Article

Sigmoidal Dependence of Electrical Conductivity of Thin PEDOT:PSS Films on Concentration of Linear Glycols as a Processing Additive

Hyeok Jo Jeong [1], Hong Jang [1], Taemin Kim [1], Taeshik Earmme [2,*] and Felix Sunjoo Kim [1,*]

[1] School of Chemical Engineering and Materials Science, Chung-Ang University, Seoul 06974, Korea; ung2ji@daum.net (H.J.J.); qwer3197@naver.com (H.J.); skxo52@naver.com (T.K.)
[2] Department of Chemical Engineering, Hongik University, Seoul 04066, Korea
* Correspondence: earmme@hongik.ac.kr (T.E.); fskim@cau.ac.kr (F.S.K.)

Abstract: We investigate the sigmoidal concentration dependence of electrical conductivity of poly(3,4-ethylenedioxythiophene):poly(styrene sulfonate) (PEDOT:PSS) processed with linear glycol-based additives such as ethylene glycol (EG), diethylene glycol (DEG), triethylene glycol (TEG), hexaethylene glycol (HEG), and ethylene glycol monomethyl ether (EGME). We observe that a sharp transition of conductivity occurs at the additive concentration of ~0.6 wt.%. EG, DEG, and TEG are effective in conductivity enhancement, showing the saturation conductivities of 271.8, 325.4, and 326.2 S/cm, respectively. Optical transmittance and photoelectron spectroscopic features are rather invariant when the glycols are used as an additive. Two different figures of merit, calculated from both sheet resistance and optical transmittance to describe the performance of the transparent electrodes, indicate that both DEG and TEG are two most effective additives among the series in fabrication of transparent electrodes based on PEDOT:PSS films with a thickness of ~50–60 nm.

Keywords: conducting polymer; PEDOT:PSS; electrical conductivity; processing additive; linear glycol; sigmoidal function

Citation: Jeong, H.J.; Jang, H.; Kim, T.; Earmme, T.; Kim, F.S. Sigmoidal Dependence of Electrical Conductivity of Thin PEDOT:PSS Films on Concentration of Linear Glycols as a Processing Additive. *Materials* **2021**, *14*, 1975. https://doi.org/10.3390/ma14081975

Academic Editor: Ju-Hyung Kim

Received: 2 March 2021
Accepted: 7 April 2021
Published: 15 April 2021

Publisher's Note: MDPI stays neutral with regard to jurisdictional claims in published maps and institutional affiliations.

1. Introduction

The poly(3,4-ethylenedioxythiophene):poly(styrene sulfonate) (PEDOT:PSS) composite has been widely studied for various applications, ranging from an active layer for electronics and energy devices to functional packaging layers [1–3]. In optoelectronic devices, PEDOT:PSS has been used as transparent and flexible electrodes, because of its high electrical conductivity, high transmittance in the visible region, high mechanical integrity, and high ruggedness in ambient conditions [4–9]. PEDOT:PSS is also an effective buffer layers for charge injection and extraction in devices [10–12]. PEDOT composites can be a redox-active component for energy storage [13,14]. The high conductivity and easy control of doping states have enabled its use as a promising p-type thermoelectric material [15–19]. Through electrochemical doping and dedoping, it is also possible to use PEDOT composites as a conductance-controllable layer in transistor [20–22]. Bioelectronic applications have also been sought from the PEDOT composites [23,24]. It should be emphasized that PEDOT and the related composites are industrially compatible because they can be easily synthesized in a large scale and processed as water-based dispersions.

The electrical conductivity of PEDOT:PSS has been largely increased by varying a processing protocol in an aqueous dispersion or by applying post-deposition treatment on thin films [5–7,25–34]. For example, diverse classes of solvents and chemicals have been used as an additive for enhancing the electrical conductivity of PEDOT:PSS films. These additives have often been called as a secondary dopants because a minute amount is added to the stock dispersion of PEDOT:PSS, although the chemicals do not seem to significantly alter the charge-carrier density. The most noticeable chemicals include dimethyl sulfoxide (DMSO) and ethylene glycol (EG) [4,30–34]. DMSO and EG have been proven to enhance the conductivity and widely used for fabrication of high-conductivity electrodes. Sorbitol, which is a sugar alcohol and in a solid form at room temperature, has also been added into PEDOT:PSS for conductivity enhancement [34,35]. Surfactants have also been mentioned as an effective additive [36–38]. Polymers with a common building block of ethylene oxide have been mixed with PEDOT:PSS dispersions and showed a positive effect on the conductivity enhancement [39–41]. The governing mechanisms of conductivity enhancement have been proposed to be phase segregation between conductive parts (i.e., PEDOT) and insulating parts (i.e., PSS), bond-structural changes of PEDOT, crystallization of PEDOT, and elimination of PSS [4,26–28,33].

In addition to the use of additives, other methods have been implemented to increase the electrical conductivity of PEDOT:PSS composites. Dipping of PEDOT:PSS films in a solvent can remove excessive PSS and increase the conductivity [8,29–32]. Treatment with sulfuric acid (H_2SO_4) and organic acids has showed to be very effective in removing PSS, separating phases, and enhancing crystallinity, resulting in the electrical conductivity as high as ~4 kS/cm [6,7,42,43]. Despite the effectiveness in conductivity enhancement, it can be also challenging to apply the post-deposition treatment because the procedure may damage the underlying layers and structures.

In this work, we investigated the changes in the electrical conductivity of PEDOT:PSS films as a function of the concentrations of various linear glycols as an additive and extracted the threshold concentrations from their sigmoidal dependencies. We chose ethylene glycol (EG), diethylene glycol (DEG), triethylene glycol (TEG), hexaethylene glycol (HEG), and ethylene glycol monomethyl ether (EGME, also known as 2-methoxyethanol) as the additive, because these share common glycol structures. Although some of these chemicals have been studied before, we have focused on their concentration dependencies of systematically varied molecular structures to quantify the transition points. The electrical conductivity of the PEDOT:PSS with additives followed the sigmoidal dependence with an inflection point at the glycol concentration of ~0.6 wt.%. We then studied the optical and photoelectron spectroscopic features of the PEDOT:PSS films processed with additives and correlated the electrical properties. We also calculated two different figures of merit, which have been often used to describe the performance of the transparent electrodes, using both sheet resistance and optical transmittance of the PEDOT:PSS films. We found that both DEG and TEG could be effective in fabrication of transparent conducting electrodes based on PEDOT:PSS films with a thickness of ~50–60 nm.

2. Materials and Methods

An aqueous dispersion of PEDOT:PSS has the concentration of 10.2 mg/mL and the nominal weight ratio of PEDOT to PSS of 1:2.5. The PEDOT:PSS dispersion was mixed with various additives at controlled molar concentrations and stirred for 24 h at room temperature. PEDOT:PSS films were prepared on a clean glass or silicon substrate by spin-coating at 3 krpm for 60 s and annealed on a hot-plate at 140 °C for 10 min in air. The absorption spectra of the films were measured using a Jasco V-670 UV-Vis-NIR spectrophotometer. The sheet resistances (R_{sh}) of the films were measured using 4-point probes in a colinear arrangement with a spacing of 1 mm and an Agilent 34450A digital multimeter. The resistance (R) in a unit of Ω, obtained by dividing the voltage difference in the inner probes by the current flowing between the outer probes, was converted to the sheet resistance (R_{sh}) in a unit of Ω per square (Ω/sq.) by multiplying a correction factor ($\pi/\ln2$) for very thin films [44]. The film thickness (d) was measured by using an Alpha-step 500 surface profilometer. The electrical conductivity (σ_{dc}) was calculated with the sheet resistance and thickness of PEDOT:PSS films ($\sigma_{dc} = 1/R_{sh}d$). At least four different points were tested to get the average values. X-ray photoelectron spectra (XPS) were obtained under ultra-high vacuum (7×10^{-9} mbar) by using Sigma Probe (Thermo VG Scientific, East Grinstead, UK) with a monochromatic Al-Kα X-ray at 15 kV and 100 W. Survey scan of XPS was performed at 50 eV for pass energy and 1.0 eV for step size. A high-resolution scan was performed at 20 eV for pass energy and 0.1 eV for step size. The XPS peaks were fitted with the Avantage program and calibrated with C1s (284.5 eV) as reference. Scanning electron microscopic imaging was conducted using S-3400N (Hitachi, Tokyo, Japan).

3. Results and Discussion

Toward the investigation of increased electrical conductivity of PEDOT:PSS film by using additives, various additives with different chemical structures have been chosen and applied. The electrical conductivity of a solidified PEDOT:PSS film is known to be increased when the stock PEDOT:PSS in a stage of aqueous dispersion is mixed with various additives such as EG, DEG, DMSO, and sorbitol [4,25–28,30,32–34,39]. Especially, EG is a commonly used agent for the conductivity enhancement of PEDOT:PSS since it is a liquid form at ambient conditions and is a mass-produced commercial chemical. To understand the compositional dependence of the electrical conductivity and other properties of PEDOT:PSS films on the structures of additives, we have applied EG and other linear glycol-type additives as the conductivity enhancing agents for PEDOT:PSS complexes (Table 1). These additives share common repeating units of ethylene oxide and chain ends of either hydroxyl or methoxy groups. These additives can be well mixed with water due to their hydrophilic moieties, and thus a homogeneous dispersion of PEDOT:PSS and spin-cast high-quality films can be obtained (Figure S1).

Table 1. Chemical structures, boiling points, and density of the additives.

Additive	Structure	Boiling Point (°C)	Density (g/cm^3)
ethylene glycol (EG)	HO–CH$_2$CH$_2$–OH	196	1.113
diethylene glycol (DEG)	HO–CH$_2$CH$_2$–O–CH$_2$CH$_2$–OH	245	1.118
triethylene glycol (TEG)	HO–[CH$_2$CH$_2$–O]$_3$–H	285 (4 mmHg)	1.124
hexaethylene glycol (HEG)	HO–[CH$_2$CH$_2$–O]$_6$–H	217 (4 mmHg)	1.127
ethylene glycol monomethyl ether (EGME)	HO–CH$_2$CH$_2$–O–CH$_3$	124	0.965

We characterized the electrical properties of the PEDOT:PSS films, which were solution-deposited after mixing with the additives at different concentrations (Figure 1). The sheet resistances (R_{sh}) of PEDOT:PSS films were measured using a colinear 4-point-probe method. In this simple and straightforward method, the contact resistances at the film/probe interfaces can be excluded by recording the voltage difference between two inner probes under a current applied between two outer probes. The typical sheet resistance of the pristine PEDOT:PSS film was ~70–120 kΩ/sq. Figure 1a shows that the sheet resistances of the PEDOT:PSS films depend on the processing conditions. The sheet resistance of the films was clearly reduced when linear glycols were added, which also agrees to the previous studies (See Table S1) [4,22,28,32–34,39–41]. Even at a low concentration of 0.1 M of TEG, the sheet resistances dropped to 565 Ω/sq and eventually stabilized at higher concentrations. EG and DEG showed the sheet resistance transition points at slightly higher concentrations of 0.3 and 0.5 M, respectively, and the R_{sh} values were stabilized at 550–600 Ω/sq at higher concentrations of additives. A much longer glycol, HEG, was also effective in reducing the sheet resistance, although the stabilized sheet resistance was 2 kΩ/sq. On the other hand, EGME, which has one hydroxyl group and one methoxy group at the ends, is relatively ineffective in the resistance reduction. The sheet resistance of EGME-added PEDOT:PSS was 20 kΩ/sq.

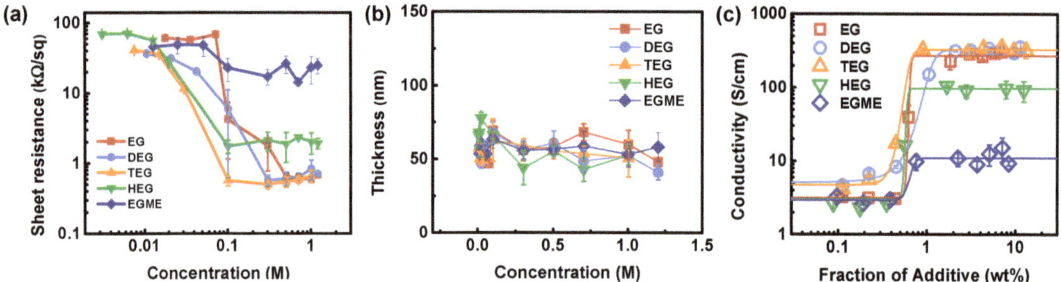

Figure 1. (**a**) Sheet resistance of poly(3,4-ethylenedioxythiophene):poly(styrene sulfonate) (PEDOT:PSS) films processed with linear glycol-based additives at various concentrations. (**b**) Film thickness of PEDOT:PSS films processed with the additives. (**c**) Electrical conductivity of PEDOT:PSS films as a function of the weight fraction of additives in the stock PEDOT:PSS dispersion. Solid lines in (**c**) are fitting curves of data with a sigmoidal logistic function as shown in Equation (1).

The sheet resistance of a conductive film was also studied with the different film thickness. Figure 1b compares the thickness of the PEDOT:PSS films processed using different concentrations of the additives. All additives decreased the film thickness, compared to the average thickness of 77.7 nm of the pristine PEDOT:PSS. The thickness values are estimated around 55–60 nm at the concentration of 1.0 M. The decrease in the film thickness is similar to the previous report on the thickness of PEDOT:PSS processed with EG as an additive [32]. The decrease in thickness when processed with additives might originate from the liquid-phase additives acting as a diluting agent during the spin-coating procedure.

From the sheet resistance and thickness of the PEDOT:PSS films, we can obtain the electrical conductivity as shown in Figure 1c. The electrical conductivity represents an intrinsic property of the materials, although the sheet resistance is an important parameter of conductive films for applications in transparent electrodes. The electrical conductivity of the PEDOT:PSS films followed a sigmoidal curve as a function of the weight fraction of additives with an inflection point. The sigmoidal dependencies have been previously reported with DEG, EG, DMSO, and sorbitol, and the threshold points have been assigned to be ~0.3–0.6 wt.% [28,34]. The sigmoidal curve is observed when a cumulative effect of a certain probability distribution is important. In the case of the conductivity enhancement in PEDOT:PSS with additives, the sigmoidal behavior can be explained phenomenologically by the probability of molecular interaction between PEDOT:PSS and the additive. Then the concentrations of both polymer and additive are the determining factors for the conductivity enhancement. To quantitatively analyze the data, we have fit the data with a sigmoidal logistic function, which is a cumulative distribution function of the logistic distribution, as in Equation (1):

$$\sigma_{dc} = \sigma_0 + \frac{\sigma_{sat} - \sigma_0}{1 + \exp\{-k(w_{add} - w_i)\}},\tag{1}$$

where σ_{dc} is the electrical conductivity of the sample in a unit of S/cm, σ_0 is the base conductivity of untreated PEDOT:PSS, σ_{sat} is the saturation conductivity, w_{add} is the weight fraction of additives in a unit of wt.% in the PEDOT:PSS dispersion, w_i is a location parameter indicating the inflection point, and k is an inverse of a scale parameter describing the steepness of the curve. Table 2 presents the fitting results of the data with the logistic function (Equation (1)), with the coefficients of determination (R^2) ranging from 0.911 for EGME to 0.997 for DEG and TEG. We note that other types of sigmoidal curves (e.g., the cumulative distribution function of the normal distribution, which can be expressed with an error function) can also describe the sigmoidal behavior. In this case, the parameters are similar to the values in Table 2 with the differences less than 7%.

Table 2. Fitting parameters of a sigmoidal logistic function (Equation (1)).

Additive	σ_{sat} (S/cm)	σ_0 (S/cm)	w_i (wt%)	k
EG	271.8	3.2	0.64	67.7
DEG	325.4	4.9	1.06	7.2
TEG	326.2	4.7	0.60	20.1
HEG	95.9	3.0	0.58	67.3
EGME	11.0	2.9	0.66	20.7

The saturation conductivities are 271.8, 325.4, and 326.2 S/cm for EG, DEG, and TEG, respectively. These polar solvents can strongly interact with PSS from the PEDOT:PSS complexes, resulting in an effective phase separation in the films and/or elimination of the insulating components from the films [26,28,29,39,45]. The saturation values are much lower for HEG (95.9 S/cm) and EGME (11.0 S/cm). We believe that the low conductivity of HEG-added PEDOT:PSS films is due to the residual additives remaining in the films. It has been previously reported that the residual solvents can limit an efficient charge transport in the polymer complexes [39]. The case of EGME is interesting, considering that the structural difference between EG and EGME is one end group (hydroxyl group vs. methoxy group). Although EGME is also polar and can enhance the conductivity, the affinity with PSS is not strong enough to result in a large degree of changes. The sheet resistance of EGME-processed PEDOT:PSS films was reduced by a factor of only ~2.4 although the post-treatment was very effective in the resistance reduction by a factor of ~200 in the previous reports [5,30]. The PEDOT:PSS films have similar transition points (w_i) of 0.6–1.0 wt.% of the linear glycol series as additives, which show close agreement to the values in the previous reports [28,34]. The coefficient k, representing how steep the curve is near the point of inflection, is the smallest for DEG. On the other hand, EG and HEG showed sharp increase in the conductivity as the additive concentration passed through the transition point. In these cases, an additive concentration of 0.8 wt.% is enough to show a saturation in the conductivity.

Such a sigmoidal dependence of the electrical conductivity on the amount of glycol additives suggests that a fraction of the components in PEDOT:PSS are influenced by the additives. The probability density function of a logistic distribution shows a symmetric curve peaked at the location parameter (i.e., w_i in Equation (1)). In this case, w_i may reflect the quantity of additives with the highest probability of interaction between the additive molecule and PEDOT:PSS. With a small amount of additive ($w_{add} < w_i$), only a small fraction of the components in PEDOT:PSS, either in an unbound form or in a complexed form, are affected and participate in molecular redistribution and reorganization. The conductivity sharply increased as the weight fraction of additives increased to near the inflection point ($w_{add} \sim w_i$) because many additive molecules can interact with PEDOT and PSS. When the fraction of additives increased further to the saturation points ($w_{add} > w_i$), there are only little molecules left to interact with. The points of inflection and saturation depend on the polymer compositions, the molecular weights, the polymerization conditions, and the degree of complexation, as well as the structure of additives.

We investigated the optical properties of PEDOT:PSS films using UV-Vis-NIR spectrophotometer, since transmittance is a critical factor for applications in transparent electrodes and the spectra can also provide information on the doping states of the PEDOT composites (Figure 2) [3,8,22,46,47]. Transmittance (T) spectra of the PEDOT:PSS films made from dispersions with the additives (1 M), regardless of the structural variations, showed a fairly high optical transparency of $T > 90\%$ in the visible region with a peak value at around 450 nm. These films can be used for transparent coating layers that pass through the visible light. The transmittance T at 550 nm, which is often used as a benchmark condition for visually transparent coatings, was >97% obtained by addition of DEG. TEGME and EGDME follow close second and third ones, respectively.

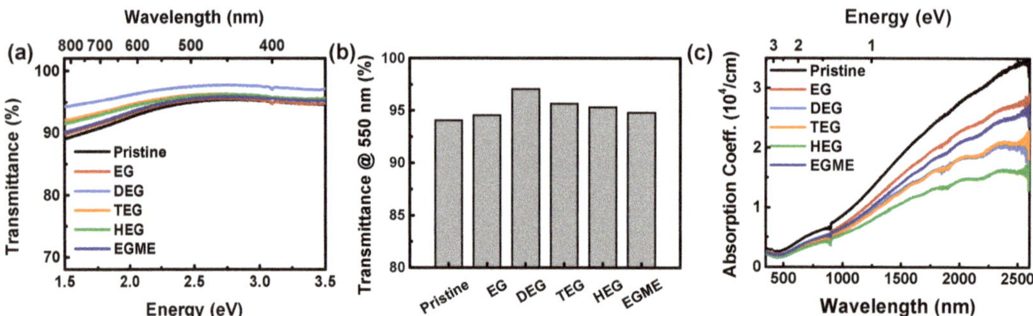

Figure 2. (**a**) Transmittance spectra, (**b**) representative transmittance values at 550 nm, and (**c**) absorption coefficients of the PEDOT:PSS films processed with various solvent additives at the concentrations of 1 M.

Typical PEDOT:PSS films show a large absorption in the near-infrared (NIR) region and relatively high transmission in the visible region. Indeed, the absorption coefficient (α) in the NIR region reaches ~10^4 cm^{-1}, while those in the visible region were an order of magnitude lower (Figure 2c). When the additives were mixed in the PEDOT:PSS dispersion, the absorbance of films slightly decreased compared to the pristine PEDOT:PSS. However, there was no noticeable change in the shape of the absorption spectra. Although the absorption coefficients vary depending on the additives, the ratios between the absorbance at NIR to that at visible regions remain similar, suggesting that the doping states are rather invariant.

From the sulfur 2p signals of X-ray photoelectron spectroscopy (XPS), we extracted the molar ratios of sulfonate to EDOT and sodium sulfonate to sulfonic acid (Figure 3). The XPS signals for S 2p at 163–170 eV in the PEDOT:PSS films should be deconvoluted, because: (a) there are multiple peaks for the aromatic S in PEDOT with a binding energy peak at 163–166 eV and the sulfonate form of PSS at 167–170 eV, (b) each of S 2p spectra shows spin-split signals (i.e., S $2p_{1/2}$ and S $2p_{3/2}$) with the corresponding ratio of 1:2, (c) the sulfur in PEDOT can be either in a neutral form (S) or in a cationic form (S$^+$), and (d) the sulfonic acid (i.e., PSSH) and sodium sulfonate (i.e., PSSNa) also have a difference of ~0.4 eV in the binding energy [26,45,48,49]. Changes in the ratios, examined by XPS, have been correlated with the conductivity increase [32,45,50]. In our samples, we found that the sulfonate-to-EDOT ratio was 2.4 for pristine PEDOT:PSS films, similar to the nominal ratio of the stock dispersion. The ratios slightly decreased to 2.29, 2.18, 2.27, 2.31, and 2.16 for EG, DEG, TEG, HEG, and EGME, respectively, as shown in Figure 3g. Because the ratios are rather close to each other, it is not clear at this moment if a change in the ratio at the surface of PEDOT:PSS films by additives is a necessary condition or not for the conductivity increase. The PSSNa/PSSH ratios dropped from 4.3 for the pristine sample to 1.6 for EG, 1.7 for DEG, 1.2 for TEG, 1.4 for HEG, and 1.9 for EGME, as shown in Figure 3h. Sodium ions can be eliminated by attractive interaction with the oxygen-rich additives.

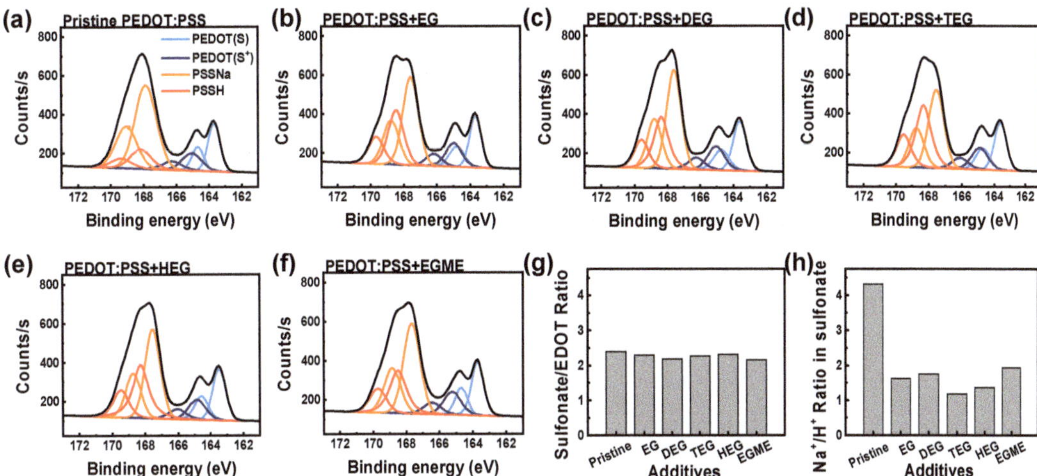

Figure 3. XPS S(2p) spectra of PEDOT:PSS films processed with various additives at 1 M: (**a**) Without additive and with (**b**) EG, (**c**) DEG, (**d**) TEG, (**e**) HEG, and (**f**) EGME. (**g**) Ratios of sulfonate to EDOT monomer using the areal ratios of deconvoluted peaks of the XPS signals. (**h**) Ratios of PSSNa/PSSH.

We investigated, parameter wise, how effective these additives are for applications in transparent conducting films by comparing two representative figures of merit. Sheet resistance (R_{sh}) and transmittance (T) of transparent conducting films are two important parameters and often combined to evaluate the figures of merit of transparent electrodes. One method has been proposed to use Equation (2):

$$\phi_{Haacke} = \frac{T^{10}}{R_{sh}} = \sigma_{dc}d \, \exp(-10\alpha d),$$ (2)

to evaluate a figure of merit in a unit of Ω^{-1} and to quantitatively compare the performance of transparent electrodes [51]. Alternatively, the two parameters, T and R_{sh}, are often correlated, at least for a thin metallic film placed in a free space, with Equation (3):

$$T = \left(1 + \frac{Z_0}{2R_{sh}} \frac{\sigma_{op}}{\sigma_{dc}}\right)^{-1},$$ (3)

where σ_{op} is the optical conductivity and Z_0 is the impedance of free space ($Z_0 = 4\pi/c = 4.19 \times 10^{-11}$ s/cm in cgs unit, which corresponds to 337 Ω in SI unit) [46,52]. In this case, a unitless ratio of σ_{dc}/σ_{op} is often used as an alternative figure of merit of transparent conductive films. It has been suggested that the ratio of >10 is required for touch panels [53]. We applied these approaches to compare the effectiveness of the additives (Figure 4). We selected the wavelength of 550 nm for comparison and found that DEG and TEG are equally effective as an additive to result in the PEDOT:PSS-based transparent electrodes. The Haacke figures of merit (Φ_{Haacke}) of the PEDOT:PSS films, calculated from Equation (2), are 1.6×10^{-3} Ω^{-1} for DEG and 2.0×10^{-3} Ω^{-1} for TEG. EG has 1.2×10^{-3} Ω^{-1}, marking the third place among the additives of interest. For the ratios of σ_{dc}/σ_{op} shown in Equation (3), we get the 15.3, 32.3, and 27.3 for EG, DEG, and TEG, respectively. As expected from the electrical conductivity of the polymer films, HEG and EGME are rather ineffective as an additive for PEDOT:PSS. We note that the DEG- and TEG-added PEDOT:PSS films are more stable than the pristine PEDOT:PSS, as the films do not show significant changes in the electrical and optical properties after aging in air for several days (Figure S2).

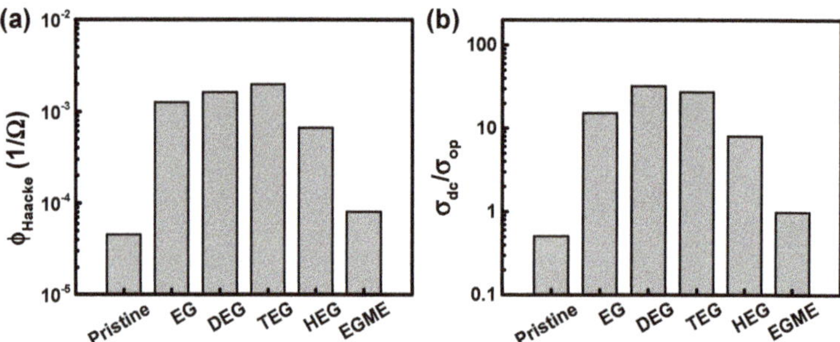

Figure 4. Figures of merit of the PEDOT:PSS films processed with additives. (**a**) The parameter proposed by Haacke. (**b**) The ratios of electrical conductivity to optical conductivity, at the wavelength of 550 nm.

4. Conclusions

In this study, we investigated the effects of linear glycols and a derivative (EG, DEG, TEG, and HEG, and EGME) as a processing additive on the electrical and optical properties of PEDOT:PSS films. We found that repeated ethylene oxides with hydroxyl groups were effective in enhancing the electrical conductivity from 3–5 S/cm in a pristine PEDOT:PSS to 270–330 S/cm with additives. We varied the concentration of the additives in the stock dispersion of PEDOT:PSS and observed sigmoidal dependence of the electrical conductivity. Through a curve fitting of the experimental conductivity–concentration data with a sigmoidal logistic function, sharp transitions were observed at the concentration of ~0.6 wt%. The PEDOT:PSS films with a thickness of 50–60 nm showed the reasonable transmittance of 94–97%. We also evaluated two different figures of merit for transparent electrodes and found that DEG and TEG among the series were effective in enhancing the performance of the PEDOT:PSS-based transparent conducting films.

Supplementary Materials: The following are available online at https://www.mdpi.com/article/10.3390/ma14081975/s1. Figure S1: SEM images of the PEDOT:PSS films. Figure S2: Relative changes in the properties after aging in air. Table S1: Comparison of the film properties with previous reports.

Author Contributions: Investigation, H.J.J., H.J., T.K., T.E. and F.S.K.; writing—original draft preparation, H.J.J. and H.J.; writing—review and editing, H.J.J., H.J., T.K., T.E. and F.S.K.; All authors have read and agreed to the published version of the manuscript.

Funding: This work was supported through the National Research Foundation of Korea (NRF) funded by the Ministry of Science and ICT (NRF-2014M3A7B4051749). This work was also supported by the Chung-Ang University Research Grants in 2019.

Data Availability Statement: The data presented in this study are available on request from the corresponding author.

Acknowledgments: We acknowledge supports from the National Research Foundation of Korea (NRF-2014M3A7B4051749) and the Chung-Ang University Research Grants in 2019.

Conflicts of Interest: The authors declare no conflict of interest.

References

1. Groenendaal, L.; Jonas, F.; Freitag, D.; Pielartzik, H.; Reynolds, J.R. Poly(3,4-ethylenedioxythiophene) and Its Derivatives: Past, Present, and Future. *Adv. Mater.* **2000**, *12*, 481–494. [CrossRef]
2. Kirchmeyer, S.; Reuter, K. Scientific importance, properties and growing applications of poly(3,4-ethylenedioxythiophene). *J. Mater. Chem.* **2005**, *15*, 2077–2088. [CrossRef]
3. Elschner, A.; Lövenich, W. Solution-deposited PEDOT for transparent conductive applications. *MRS Bull.* **2011**, *36*, 794–798. [CrossRef]
4. Ouyang, J.; Chu, C.W.; Chen, F.C.; Xu, Q.; Yang, Y. High-Conductivity Poly(3,4-ethylenedioxythiophene):Poly(styrene sulfonate) Film and Its Application in Polymer Optoelectronic Devices. *Adv. Funct. Mater.* **2005**, *15*, 203–208. [CrossRef]

5. Hsiao, Y.-S.; Whang, W.-T.; Chen, C.-P.; Chen, Y.-C. High-conductivity poly(3,4-ethylenedioxythiophene):poly(styrene sulfonate) film for use in ITO-free polymer solar cells. *J. Mater. Chem.* **2008**, *18*, 5948–5955. [CrossRef]
6. Kim, N.; Kee, S.; Lee, S.H.; Lee, B.H.; Kahng, Y.H.; Jo, Y.-R.; Kim, B.-J.; Lee, K. Highly Conductive PEDOT:PSS Nanofibrils Induced by Solution-Processed Crystallization. *Adv. Mater.* **2014**, *26*, 2268–2272. [CrossRef]
7. Kim, S.; Sanyoto, B.; Park, W.-T.; Kim, S.; Mandal, S.; Lim, J.-C.; Noh, Y.-Y.; Kim, J.-H. Purification of PEDOT:PSS by Ultrafiltration for Highly Conductive Transparent Electrode of All-Printed Organic Devices. *Adv. Mater.* **2016**, *28*, 10149–10154. [CrossRef]
8. Jang, H.; Kim, M.S.; Jang, W.; Son, H.; Wang, D.H.; Kim, F.S. Highly conductive PEDOT:PSS electrode obtained via post-treatment with alcoholic solvent for ITO-free organic solar cells. *J. Ind. Eng. Chem.* **2020**, *86*, 205–210. [CrossRef]
9. Jeong, S.-H.; Ahn, S.; Lee, T.-W. Strategies to Improve Electrical and Electronic Properties of PEDOT:PSS for Organic and Perovskite Optoelectronic Devices. *Macromol. Res.* **2019**, *27*, 2–9. [CrossRef]
10. Earmme, T.; Jenekhe, S.A. Solution-Processed, Alkali Metal-Salt-Doped, Electron-Transport Layers for High-Performance Phosphorescent Organic Light-Emitting Diodes. *Adv. Funct. Mater.* **2012**, *22*, 5126–5136. [CrossRef]
11. Cha, H.; Li, J.; Li, Y.; Kim, S.-O.; Kim, Y.-H.; Kwon, S.-K. Effects of Bulk Heterojunction Morphology Control via Thermal Annealing on the Fill Factor of Anthracene-based Polymer Solar Cells. *Macromol. Res.* **2020**, *28*, 820–825. [CrossRef]
12. Wang, L.; Park, J.S.; Lee, H.G.; Kim, G.-U.; Kim, D.; Kim, C.; Lee, S.; Kim, F.S.; Kim, B.J. Impact of Chlorination Patterns of Naphthalenediimide-Based Polymers on Aggregated Structure, Crystallinity, and Device Performance of All-Polymer Solar Cells and Organic Transistors. *ACS Appl. Mater. Interfaces* **2020**, *12*, 56240–56250. [CrossRef]
13. Österholm, A.M.; Shen, D.E.; Dyer, A.L.; Reynolds, J.R. Optimization of PEDOT Films in Ionic Liquid Supercapacitors: Demonstration As a Power Source for Polymer Electrochromic Devices. *ACS Appl. Mater. Interfaces* **2013**, *5*, 13432–13440. [CrossRef]
14. Gulercan, D.; Commandeur, D.; Chen, Q.; Sarac, A.S. A Ternary PEDOT-TiO2-Reduced Graphene Oxide Nanocomposite for Supercapacitor Applications. *Macromol. Res.* **2019**, *27*, 867–875. [CrossRef]
15. Bubnova, O.; Khan, Z.U.; Malti, A.; Braun, S.; Fahlman, M.; Berggren, M.; Crispin, X. Optimization of the thermoelectric figure of merit in the conducting polymer poly(3,4-ethylenedioxythiophene). *Nat. Mater.* **2011**, *10*, 429–433. [CrossRef] [PubMed]
16. Kim, G.H.; Shao, L.; Zhang, K.; Pipe, K.P. Engineered doping of organic semiconductors for enhanced thermoelectric efficiency. *Nat. Mater.* **2013**, *12*, 719–723. [CrossRef] [PubMed]
17. Massonnet, N.; Carella, A.; Jaudouin, O.; Rannou, P.; Laval, G.; Celle, C.; Simonato, J.-P. Improvement of the Seebeck coefficient of PEDOT:PSS by chemical reduction combined with a novel method for its transfer using free-standing thin films. *J. Mater. Chem. C* **2014**, *2*, 1278–1283. [CrossRef]
18. Park, H.; Lee, S.H.; Kim, F.S.; Choi, H.H.; Cheong, I.W.; Kim, J.H. Enhanced thermoelectric properties of PEDOT:PSS nanofilms by a chemical dedoping process. *J. Mater. Chem. A* **2014**, *2*, 6532–6539. [CrossRef]
19. Lee, S.; Kim, S.; Pathak, A.; Tripathi, A.; Qiao, T.; Lee, Y.; Lee, H.; Woo, H.Y. Recent Progress in Organic Thermoelectric Materials and Devices. *Macromol. Res.* **2020**, *28*, 531–552. [CrossRef]
20. Bernards, D.A.; Malliaras, G.G. Steady-State and Transient Behavior of Organic Electrochemical Transistors. *Adv. Funct. Mater.* **2007**, *17*, 3538–3544. [CrossRef]
21. Kim, S.-M.; Kim, C.-H.; Kim, Y.; Kim, N.; Lee, W.-J.; Lee, E.-H.; Kim, D.; Park, S.; Lee, K.; Rivnay, J.; et al. Influence of PEDOT:PSS crystallinity and composition on electrochemical transistor performance and long-term stability. *Nat. Commun.* **2018**, *9*, 3858. [CrossRef]
22. Kim, D.; Jang, H.; Lee, S.; Kim, B.J.; Kim, F.S. Solid-State Organic Electrolyte-Gated Transistors Based on Doping-Controlled Polymer Composites with a Confined Two-Dimensional Channel in Dry Conditions. *ACS Appl. Mater. Interfaces* **2021**, *13*, 1065–1075. [CrossRef] [PubMed]
23. Rivnay, J.; Leleux, P.; Ferro, M.; Sessolo, M.; Williamson, A.; Koutsouras, D.A.; Khodagholy, D.; Ramuz, M.; Strakosas, X.; Owens, R.M.; et al. High-performance transistors for bioelectronics through tuning of channel thickness. *Sci. Adv.* **2015**, *1*, e1400251. [CrossRef] [PubMed]
24. Harman, D.G.; Gorkin Iii, R.; Stevens, L.; Thompson, B.; Wagner, K.; Weng, B.; Chung, J.H.Y.; in het Panhuis, M.; Wallace, G.G. Poly(3,4-ethylenedioxythiophene):dextran sulfate (PEDOT:DS)–A highly processable conductive organic biopolymer. *Acta Biomater.* **2015**, *14*, 33–42. [CrossRef] [PubMed]
25. Huang, J.; Miller, P.F.; Wilson, J.S.; de Mello, A.J.; de Mello, J.C.; Bradley, D.D.C. Investigation of the Effects of Doping and Post-Deposition Treatments on the Conductivity, Morphology, and Work Function of Poly(3,4-ethylenedioxythiophene)/Poly(styrene sulfonate) Films. *Adv. Funct. Mater.* **2005**, *15*, 290–296. [CrossRef]
26. Kim, J.Y.; Jung, J.H.; Lee, D.E.; Joo, J. Enhancement of electrical conductivity of poly(3,4-ethylenedioxythiophene)/poly(4-styrenesulfonate) by a change of solvents. *Synth. Met.* **2002**, *126*, 311–316. [CrossRef]
27. Timpanaro, S.; Kemerink, M.; Touwslager, F.J.; De Kok, M.M.; Schrader, S. Morphology and conductivity of PEDOT/PSS films studied by scanning–tunneling microscopy. *Chem. Phys. Lett.* **2004**, *394*, 339–343. [CrossRef]
28. Crispin, X.; Jakobsson, F.L.E.; Crispin, A.; Grim, P.C.M.; Andersson, P.; Volodin, A.; van Haesendonck, C.; Van der Auweraer, M.; Salaneck, W.R.; Berggren, M. The Origin of the High Conductivity of Poly(3,4-ethylenedioxythiophene)−Poly(styrenesulfonate) (PEDOT−PSS) Plastic Electrodes. *Chem. Mater.* **2006**, *18*, 4354–4360. [CrossRef]
29. Ouyang, J.; Xu, Q.; Chu, C.-W.; Yang, Y.; Li, G.; Shinar, J. On the mechanism of conductivity enhancement in poly(3,4-ethylenedioxythiophene):poly(styrene sulfonate) film through solvent treatment. *Polymer* **2004**, *45*, 8443–8450. [CrossRef]

30. Martin, B.D.; Nikolov, N.; Pollack, S.K.; Saprigin, A.; Shashidhar, R.; Zhang, F.; Heiney, P.A. Hydroxylated secondary dopants for surface resistance enhancement in transparent poly(3,4-ethylenedioxythiophene)–poly(styrenesulfonate) thin films. *Synth. Met.* **2004**, *142*, 187–193. [CrossRef]

31. Kim, N.; Lee, B.H.; Choi, D.; Kim, G.; Kim, H.; Kim, J.-R.; Lee, J.; Kahng, Y.H.; Lee, K. Role of Interchain Coupling in the Metallic State of Conducting Polymers. *Phys. Rev. Lett.* **2012**, *109*, 106405. [CrossRef]

32. Kim, Y.H.; Sachse, C.; Machala, M.L.; May, C.; Müller-Meskamp, L.; Leo, K. Highly Conductive PEDOT:PSS Electrode with Optimized Solvent and Thermal Post-Treatment for ITO-Free Organic Solar Cells. *Adv. Funct. Mater.* **2011**, *21*, 1076–1081. [CrossRef]

33. Ouyang, L.; Musumeci, C.; Jafari, M.J.; Ederth, T.; Inganäs, O. Imaging the Phase Separation Between PEDOT and Polyelectrolytes During Processing of Highly Conductive PEDOT:PSS Films. *ACS Appl. Mater. Interfaces* **2015**, *7*, 19764–19773. [CrossRef]

34. Nevrela, J.; Micjan, M.; Novota, M.; Kovacova, S.; Pavuk, M.; Juhasz, P.; Kovac, J.; Jakabovic, J.; Weis, M. Secondary doping in poly(3,4-ethylenedioxythiophene):Poly(4-styrenesulfonate) thin films. *J. Polym. Sci. Part B: Polym. Phys.* **2015**, *53*, 1139–1146. [CrossRef]

35. Zhang, F.; Johansson, M.; Andersson, M.R.; Hummelen, J.C.; Inganäs, O. Polymer Photovoltaic Cells with Conducting Polymer Anodes. *Adv. Mater.* **2002**, *14*, 662–665. [CrossRef]

36. Fan, B.; Mei, X.; Ouyang, J. Significant Conductivity Enhancement of Conductive Poly(3,4-ethylenedioxythiophene):Poly(styrenesulfonate) Films by Adding Anionic Surfactants into Polymer Solution. *Macromolecules* **2008**, *41*, 5971–5973. [CrossRef]

37. Choi, J.S.; Yim, J.-H.; Kim, D.-W.; Jeon, J.-K.; Ko, Y.S.; Kim, Y. Effects of various imidazole-based weak bases and surfactant on the conductivity and transparency of poly(3,4-ethylenedioxythiophene) films. *Synth. Met.* **2009**, *159*, 2506–2511. [CrossRef]

38. Vosgueritchian, M.; Lipomi, D.J.; Bao, Z. Highly Conductive and Transparent PEDOT:PSS Films with a Fluorosurfactant for Stretchable and Flexible Transparent Electrodes. *Adv. Funct. Mater.* **2012**, *22*, 421–428. [CrossRef]

39. Mengistie, D.A.; Wang, P.-C.; Chu, C.-W. Effect of molecular weight of additives on the conductivity of PEDOT:PSS and efficiency for ITO-free organic solar cells. *J. Mater. Chem. A* **2013**, *1*, 9907–9915. [CrossRef]

40. Lee, J.J.; Lee, S.H.; Kim, F.S.; Choi, H.H.; Kim, J.H. Simultaneous enhancement of the efficiency and stability of organic solar cells using PEDOT:PSS grafted with a PEGME buffer layer. *Org. Electron.* **2015**, *26*, 191–199. [CrossRef]

41. Wang, T.; Qi, Y.; Xu, J.; Hu, X.; Chen, P. Effects of poly(ethylene glycol) on electrical conductivity of poly(3,4-ethylenedioxythiophene)-poly(styrenesulfonic acid) film. *Appl. Surf. Sci.* **2005**, *250*, 188–194. [CrossRef]

42. Xia, Y.; Sun, K.; Ouyang, J. Solution-Processed Metallic Conducting Polymer Films as Transparent Electrode of Optoelectronic Devices. *Adv. Mater.* **2012**, *24*, 2436–2440. [CrossRef] [PubMed]

43. Ouyang, J. Solution-Processed PEDOT:PSS Films with Conductivities as Indium Tin Oxide through a Treatment with Mild and Weak Organic Acids. *ACS Appl. Mater. Interfaces* **2013**, *5*, 13082–13088. [CrossRef]

44. Uhlir, A. The potentials of infinite systems of sources and numerical solutions of problems in semiconductor engineering. *Bell Syst. Tech. J.* **1955**, *34*, 105–128. [CrossRef]

45. Crispin, X.; Marciniak, S.; Osikowicz, W.; Zotti, G.; van der Gon, A.W.D.; Louwet, F.; Fahlman, M.; Groenendaal, L.; De Schryver, F.; Salaneck, W.R. Conductivity, morphology, interfacial chemistry, and stability of poly(3,4-ethylene dioxythiophene)–poly(styrene sulfonate): A photoelectron spectroscopy study. *J. Polym. Sci. Part B: Polym. Phys.* **2003**, *41*, 2561–2583. [CrossRef]

46. Ellmer, K. Past achievements and future challenges in the development of optically transparent electrodes. *Nat. Photon.* **2012**, *6*, 809–817. [CrossRef]

47. van der Pol, T.P.A.; Keene, S.T.; Saes, B.W.H.; Meskers, S.C.J.; Salleo, A.; van de Burgt, Y.; Janssen, R.A.J. The Mechanism of Dedoping PEDOT:PSS by Aliphatic Polyamines. *J. Phys. Chem. C* **2019**, *123*, 24328–24337. [CrossRef]

48. Greczynski, G.; Kugler, T.; Salaneck, W.R. Characterization of the PEDOT-PSS system by means of X-ray and ultraviolet photoelectron spectroscopy. *Thin Solid Films* **1999**, *354*, 129–135. [CrossRef]

49. Khan, M.A.; Armes, S.P.; Perruchot, C.; Ouamara, H.; Chehimi, M.M.; Greaves, S.J.; Watts, J.F. Surface characterization of poly(3,4-ethylenedioxythiophene)-coated latexes by X-ray photoelectron spectroscopy. *Langmuir* **2000**, *16*, 4171–4179. [CrossRef]

50. Cho, H.; Cho, W.; Kim, Y.; Lee, J.-G.; Kim, J.H. Influence of residual sodium ions on the structure and properties of poly(3,4-ethylenedioxythiophene):poly(styrenesulfonate). *RSC Adv.* **2018**, *8*, 29044–29050. [CrossRef]

51. Haacke, G. New figure of merit for transparent conductors. *J. Appl. Phys.* **1976**, *47*, 4086–4089. [CrossRef]

52. Dressel, M.; Grüner, G. *Electrodynamics of Solids: Optical Properties of Electrons in Matter*; Cambridge University Press: Cambridge, UK, 2002. [CrossRef]

53. Hecht, D.S.; Hu, L.; Irvin, G. Emerging Transparent Electrodes Based on Thin Films of Carbon Nanotubes, Graphene, and Metallic Nanostructures. *Adv. Mater.* **2011**, *23*, 1482–1513. [CrossRef]

 *materials*

Article

Thermogelling Behaviors of Aqueous Poly(N-Isopropylacrylamide-co-2-Hydroxyethyl Methacrylate) Microgel–Silica Nanoparticle Composite Dispersions

Byung Soo Hwang [1,†], Jong Sik Kim [2,†], Ju Min Kim [1,2,*] and Tae Soup Shim [1,2,*]

[1] Department of Chemical Engineering, Ajou University, Suwon 16499, Korea; brianify@ajou.ac.kr
[2] Department of Energy Systems Research, Ajou University, Suwon 16499, Korea; kjsik@ajou.ac.kr
* Correspondence: jumin@ajou.ac.kr (J.M.K.); tsshim@ajou.ac.kr (T.S.S.)
† These authors contributed equally to this work.

Abstract: Gelation behaviors of hydrogels have provided an outlook for the development of stimuli-responsive functional materials. Of these materials, the thermogelling behavior of poly(N-isopropylacrylamide) (p(NiPAm))-based microgels exhibits a unique, reverse sol–gel transition by bulk aggregation of microgels at the lower critical solution temperature (LCST). Despite its unique phase transition behaviors, the application of this material has been largely limited to the biomedical field, and the bulk gelation behavior of microgels in the presence of colloidal additives is still open for scrutinization. Here, we provide an in-depth investigation of the unique thermogelling behaviors of p(NiPAm)-based microgels through poly(N-isopropylacrylamide-co-2-hydroxyethyl methacrylate) microgel (p(NiPAm-co-HEMA))–silica nanoparticle composite to expand the application possibilities of the microgel system. Thermogelling behaviors of p(NiPAm-co-HEMA) microgel with different molar ratios of N-isopropylacrylamide (NiPAm) and 2-hydroxyethyl methacrylate (HEMA), their colloidal stability under various microgel concentrations, and the ionic strength of these aqueous solutions were investigated. In addition, sol–gel transition behaviors of various p(NiPAm-co-HEMA) microgel systems were compared by analyzing their rheological properties. Finally, we incorporated silica nanoparticles to the microgel system and investigated the thermogelling behaviors of the microgel–nanoparticle composite system. The composite system exhibited consistent thermogelling behaviors in moderate conditions, which was confirmed by an optical microscope. The composite demonstrated enhanced mechanical strength at gel state, which was confirmed by analyzing rheological properties.

Keywords: stimuli-responsive hydrogels; thermogelling polymers; sol–gel transition behaviors; complex colloidal systems

Citation: Hwang, B.S.; Kim, J.S.; Kim, J.M.; Shim, T.S. Thermogelling Behaviors of Aqueous Poly(N-Isopropylacrylamide-co-2-Hydroxyethyl Methacrylate) Microgel–Silica Nanoparticle Composite Dispersions. *Materials* **2021**, *14*, 1212. https://doi.org/10.3390/ma14051212

Academic Editor: Philippe Colomban

Received: 25 January 2021
Accepted: 28 February 2021
Published: 4 March 2021

Publisher's Note: MDPI stays neutral with regard to jurisdictional claims in published maps and institutional affiliations.

1. Introduction

Phase transition behaviors of colloidal dispersion have long been studied to understand the fundamentals of colloidal interactions and to apply the system to various fields such as cosmetics, pharmaceutics, food industries, paints, inks, slurries, etc. Especially, in-depth investigation of highly complex systems of slurries and pastes based on the knowledge of rheological properties of homogeneous or heterogeneous colloidal formulations paves a promising outlook for the energy industry [1,2]. However, most of the systems applied in the industry are based on hard-sphere-like colloids such as carbon, polystyrene, and silica. In comparison, investigations on soft materials such as hydrogels in the previously discussed fields have had a weaker standing.

Hydrogels are water-absorbing polymers well-known for their bio-compatibility and stimuli-responsivity. The reversible volume phase transition and elastic behaviors of hydrogel have enabled the design of smart materials such as temperature-responsive drug delivery and wound healing materials [3–5], remotely controlled soft actuators [6,7], stimuli-responsive plasmonic materials [8], flexible sensors [9,10], etc. In addition, a sparse polymer network in an aqueous medium can be used as a matrix to incorporate nanoparticles [11] and microreactors [12,13]. However, in the past, their intrinsic drawbacks such as weak mechanical strength and mandate for the moist environment have limited the field of applications. Recent achievements addressed these issues: the development of double-network hydrogels [14] which enhances the mechanical durability and replacement of the aqueous medium with a non-volatile organic medium that improves the useability in dry environments [15]. Therefore, we can expect a gradual expansion in the application field of hydrogel-based soft colloids to the field areas in which hard-sphere-like colloids are used while maintaining their functionality.

Of the hydrogels, the thermogelling microgel dispersion is one of the intriguing systems that exhibits reversible sol–gel transition behavior [16]. Unlike the conventional physical gels that maintain a gel state below the phase transition temperature, the thermogelling microgel dispersion shows inverse phase transition from sol to gel as the temperature increases. This inverse phase transition is triggered by the hydrophilic and hydrophobic interactions among microgels at the volume phase transition temperature of the material [17]. Poly(n-isopropylacrylamide) p(NiPAm) microgel dispersion is one of the widely known temperature-responsive colloid system. In general, it shows reversible swelling/shrinkage at a lower critical solution temperature (LCST), which is around 32 °C. Above LCST, the hydrophobic moiety becomes dominant and repels water out from the polymer network of the microgel [18]. The p(NiPAm) colloids show a stable dispersion even at a temperature above LCST. When salt is added, however, it forms bulk gels around LCST due to the weak electrostatic repulsion among microgels [19]. Because this gelation does not require additional crosslinking reactions, they have been studied as an in situ gelation material for biomedical applications. Furthermore, copolymerization of p(NiPAm) with various comonomers has been investigated to harness sol–gel transition behaviors. Many endeavors were put to enhance the mechanical strength of the gel while maintaining the sol–gel transition behavior, and many successes have been reported. For example, copolymerization of NiPAm with methylcellulose [20] or acrylic acid [21] showed that the gel's mechanical strength could be enhanced with varying concentrations. A different enhancement, the maintenance of the gel volume at LCST, also shared some attention. For instance, copolymerization with 2-hydroxyethyl methacrylate (HEMA) extended the sequential delivery of the protein from the gel scaffold, which can be interpreted as an improvement from subsequent shrinkage of the gel after gelation [22]. Despite many studies on understanding gelation and controlling their mechanical properties, the thermogelling behavior of p(NiPAm)-based microgels in the presence of colloidal additives has not been fully understood. To take advantage of the unique gelation behavior in various colloidal systems, it is necessary to understand the effect of colloidal additives. Following the research mentioned, investigating the phase transition behaviors of microgel–nanoparticle composite systems will prove useful to expand the areas of application to more than just the biomedical field.

In this study, we specifically consider thermogelling behaviors of microgels–nanoparticle composite systems consisted of poly(N-isopropylacrylamide-co-2-hydroxyethyl methacrylate) (p(NiPAm-co-HEMA)) microgels and silica nanoparticles. We prepared aqueous p(NiPAm-co-HEMA) microgel dispersions with various NiPAm:HEMA molar ratios through radical polymerization. Then, thermogelling behaviors of the neat p(NiPAm-co-HEMA) microgel dispersions upon varying salt concentrations were studied by a rheometer. Finally, we added silica nanoparticles to the microgel dispersion to make a model composite system to investigate the thermogelling behaviors.

2. Materials and Methods

2.1. Materials

N-Isopropylacrylamide (NiPAm, TCI, Tokyo, Japan), 2-hydroxyethyl methacrylate (HEMA, Aldrich, St. Louis, MO, USA), N,N′-methylenebis(acrylamide) (MBA, Aldrich, St. Louis, MO, USA), ammonium persulfate (APS, Aldrich, St. Louis, MO, USA), and sodium dodecyl sulfate (SDS, Aldrich, St. Louis, MO, USA) were purchased and used as received.

2.2. Synthesis of p(NiPAm-co-HEMA) Microgels

Microgels of the different molar ratios of NiPAm and HEMA were synthesized by free radical precipitation polymerization. The constituent monomers, NiPAm and HEMA, and the crosslinker, MBA, were dissolved accordingly to the specified molar ratio while the amount of the initiator, APS, and the surfactant, SDS, were kept constant. Six batches of p(NiPAm-co-HEMA) were synthesized, namely, 10:0, 9:1, 8:2, 7:3, 6:4, 5:5, denoting the molar ratio between NiPAm and HEMA. The detailed recipe of each batch is summarized in Table 1. The monomers, crosslinker, and surfactant were dissolved in 498 mL of water. The reaction mixture was transferred to a 1 L-size, three-neck, and double-jacket reactor equipped with a condenser, a nitrogen inlet, and a mechanical stirrer. The reaction mixture was stabilized at 70 °C while stirring at 360 rpm under a nitrogen atmosphere for an hour. Afterward, 0.342 g of APS was dissolved in 2 mL of water and added to the mixture to initiate the polymerization. The reaction time for each batch is specified in Table 1. The resultant microgel dispersion was washed with three successive centrifugations to remove unreacted monomers and the surfactant. Finally, the dispersion was freeze-dried to obtain a solid sample for future use.

Table 1. Reaction recipe for the synthesis of poly(N-isopropylacrylamide-co-2-hydroxyethyl methacrylate) (p(NiPAm-co-HEMA)) microgels.

NiPAm: HEMA Molar Ratio	NiPAm		HEMA		MBA		SDS		APS		Reaction Time
	mmol	g	mmol	g	mmol	g	mmol	g	mmol	g	h
10:0	250	28.290	0.00	0.000							1
9:1	225	25.461	25.0	3.254							1
8:2	200	22.632	50.0	6.507	5.00	0.771					1
7:3	175	19.803	75.0	9.761			1.00	0.288	1.50	0.342	1
6:4	150	16.974	100	13.014							0.75
5:5	75	8.487	75	9.761	3.00	0.463					0.75

2.3. Rheological Characterization of p(NiPAm-co-HEMA) Microgels

The gelling behavior of the p(NiPAm-co-HEMA) microgels was characterized by small-amplitude oscillatory shear (SAOS) tests. We used a commercial rotational rheometer (AR-G2, TA Instruments, New Castle, DE, USA) equipped with a cone-and-plate geometry (1°; diameter: 60 mm) to perform the linear viscoelasticity characterization. The storage and loss moduli (G′ and G″) were monitored at the fixed frequency of 0.63 (rad/s) with increasing temperature starting from 20 °C with the increment rate at 1 °C/min.

3. Results and Discussion

3.1. Synthesis and Characterization of p(NiPAm-co-HEMA) Microgels

The p(NiPAm-co-HEMA) microgels were synthesized by radical precipitation polymerization as illustrated in Figure 1a [23]. We simply mixed two monomers having unsaturated hydrocarbon groups with a crosslinker. Thus, it is assumed that the polymeric structure of the copolymer is random. For Fourier-transform infrared spectroscopy (FTIR) measurement, p(NiPAm-co-HEMA) microgel was freeze-dried to obtain a powdery sample. The FTIR data in Figure 1b confirms the successful synthesis of p(NiPAm-co-HEMA) in accordance with the specified mole ratio as described in the experimental section. The absorption peak at 3275 cm^{-1} and 1640 cm^{-1} are assigned to stretching vibration of the amino group (N–H) and the amide I groups (C=O) in NiPAM, respectively [24]. The peak at 1734 cm^{-1} is assigned to the stretching vibration of the carbonyl group (C=O) in HEMA [17]. As the molar ratio of HEMA increases so does the peak intensity at 1734 cm^{-1}, while it dwindles at 1640 cm^{-1}. Normalized FTIR data for comparison of NiPAm and HEMA peak ratio of p(NiPAm-co-HEMA) microgels showed an increase of carbonyl group signal as the content of HEMA increased. For the 5:5 sample, however, the intensity of the carbonyl group signal decreased because the microgel was not successfully synthesized. Zeta potential of microgel dispersion with varying NiPAm:HEMA molar ratio was measured at room temperature. For a control sample, p(NiPAm) microgel dispersion, the value of zeta potential was at the isoelectric point with a slightly negative value. It showed a decreasing trend in values as the content of HEMA increased.

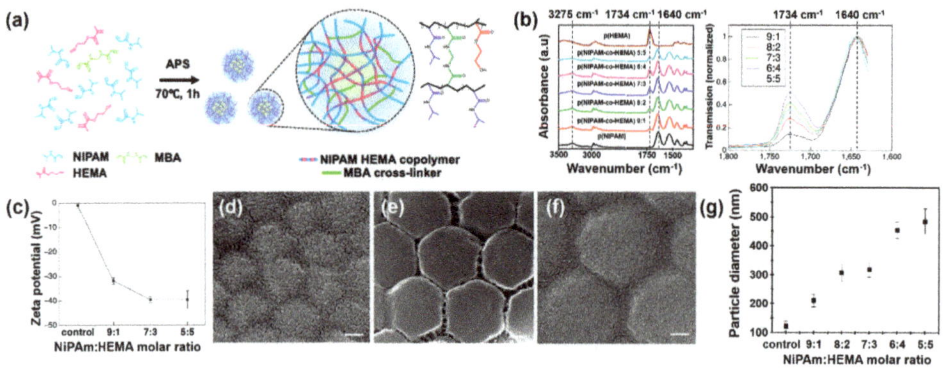

Figure 1. (**a**) Schematic illustration for the synthesis of p(NiPAm-co-HEMA) microgels. (**b**) Fourier-transform infrared spectroscopy (FTIR) data for p(HEMA) polymer, p(NiPAm), and p(NiPAm-co-HEMA) microgels with varying molar ratio between N-isopropylacrylamide (NiPAm) and 2-hydroxyethyl methacrylate (HEMA). For comparison of NiPAm and HEMA signal ratio among samples, normalized FTIR data are plotted for p(NiPAm-co-HEMA) microgels (right). (**c**) Zeta potential of p(NiPAm) and p(NiPAm-co-HEMA) microgels with varying NiPAm:HEMA molar ratio. (**d–f**) Scanning electron microscope (SEM) images of p(NiPAm-co-HEMA) microgels with molar ratios between NiPAm and HEMA of (**d**) 9:1, (**e**) 7:3, and (**f**) 5:5, respectively. Scale bar in (**d–f**) represents 100 nm. (**g**) Effect of molar composition of HEMA to particle diameter.

The average diameter of p(NiPAm-co-HEMA) microgels shows a degree of swelling. Scanning electron microscope (SEM) images of the freeze-dried p(NiPAm-co-HEMA) microgels in Figure 1d–f indicate uniformly sized, spherical microgels. It was confirmed that the diameter of microgels increased as the composition of HEMA increased, Figure 1g. The increment of particle size is attributed to the good water uptake capability of HEMA. As a result, sparse polymeric networks were formed during the radical polymerization as the amount of HEMA increased, resulting in the formation of larger-, and mechanically weaker microgels. Indeed, when we increased the molar ratio of HEMA to more than 0.5, the structural integrity of p(NIPAM-co-HEMA) microgels could not be maintained. As a result, bulk p(NIPAM-co-HEMA) polymer cakes were observed as gels precipitated. Based on the above results, further studies regarding investigations of colloidal stability and their thermogelling behaviors were mainly focused on the characteristics of p(NiPAm-HEMA) microgel dispersions with NiPAm to HEMA molar ratios of 9:1, 7:3, and 5:5. In addition, further experiments were conducted in neutral pH conditions because colloidal stability under acidic and alkalic conditions was unstable, as shown in Figure S1.

3.2. Effect of Salt on the Stability of p(NiPAm-co-HEMA) Microgel Dispersion

A notable difference between p(NiPAm-co-HEMA) and p(NiPAm) microgel dispersion is that the former requires the addition of salt and appropriate temperature for successful gelation, Figure 2a. This is because the electrostatic repulsion between the microgels is increased by the hydroxyl group of HEMA. In the presence of HEMA, salt ions are required to screen the electrostatic double layers on the microgels to prompt the formation of a particle network that leads to gelation upon thermal heating. Because adding salt in conventional colloidal dispersion causes destabilization of the system, we first investigated the effect of salt on the stability of p(NiPAm-co-HEMA) microgel dispersions.

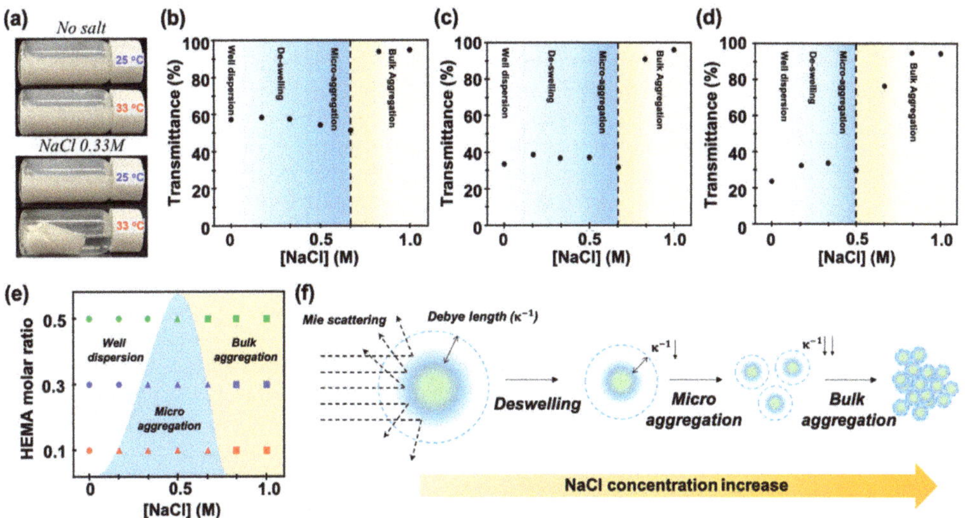

Figure 2. (**a**) Photograph images of thermogelling p(NiPAm-co-HEMA) microgel dispersion upon different temperature conditions. In the absence of salt, the dispersion remained at the sol state for both 25 °C and 33 °C (top panel). In the presence of 0.33 M of sodium chloride (NaCl), microgels dispersion showed volume phase transition from sol to gel state at 33 °C (bottom panel). (**b–d**) The transmittance of p(NiPAm-co-HEMA) microgel dispersion with molar ratios between NiPAm and HEMA of (**b**) 9:1, (**c**) 7:3, and (**d**) 5:5, respectively, with respect to NaCl concentration. The bulk aggregation of microgels by the strong ionic strength of solution was expressed in yellow background. (**e**) Summary of dispersion quality of p(NiPAm-co-HEMA) microgels. The microgel showed a well-dispersed (white background) phase, micro aggregation (blue background) phase, and bulk aggregation (yellow background) phase in accordance with NiPAm: HEMA molar ratios and NaCl concentrations. (**f**) Schematic illustration of deswelling of microgel with the increase of the salt concentration.

The stability of p(NiPAm-co-HEMA) microgel dispersion was investigated by observing the turbidity of the dispersion via a UV–visible spectrometer (SHIMAZU, Kyoto, Japan). 0.04 wt% of p(NiPAm-co-HEMA) microgel dispersion with varying NiPAm to HEMA molar ratios were prepared. As shown in Figure 2b–d, the transmittance of the p(NiPAm-co-HEMA) microgels composed of NiPAm to HEMA molar ratios of 9:1, 7:3, and 5:5 at different NaCl concentrations were measured. The transmittance for each sample without NaCl was measured as 59.7%, 34.7%, and 23.7% for 9:1, 7:3, and 5:5 samples, respectively, which showed a decreasing trend as the amount of HEMA increased. This is because the particle size-dependent Mie scattering was dominant for particles with a size similar to the wavelength of the incident light. In all cases, transmittance was slightly increased at low salt concentration and then started to decrease until it reached the critical salt concentration. Then, a dramatic increase in transmittance was observed. These results can be explained by the change of Mie scattering and Debye length of the system along with the change of the salt concentration. When the salt is added to the system, osmotic pressure applied to the microgel causes the deswelling of the microgels [16,25]. It results in the decrease of the microgel size and thereby decrease of Mie scattering. Therefore, the transmittance of dispersion was increased. As we stated above, an increase of the salt concentration in colloidal dispersion also causes the screening of electric double layer, which results in the decrease of Debye length of microgels. When the Debye length reaches below the critical value for maintaining a stable dispersion, Van der Waals force causes aggregation of microgels followed by sedimentation. Here, a decrease and dramatic increase of transmittance can be explained by microaggregation and bulk aggregation (i.e., sedimentation), respectively. When the dispersion is at the state of microaggregation, the effective size of microgels can be considered to be larger and it results in the decrease of transmittance. Then, a sudden increase of transmittance at bulk aggregation is observed due to the sedimentation. Therefore, a critical salt concentration can be regarded as a point at which bulk aggregation of microgels occurs. The phase behaviors of p(NiPAm-co-HEMA) microgel dispersion with varying NiPAm:HEMA compositions in Figure 2e revealed that the higher the amount of HEMA is, the more the salt concentration-initiated microaggregation occurs. By reflecting on the trends of transmittance for each sample, it is expected that the content of HEMA contributed to the increase of Debye length of microgels, which also agreed with the previous results. The comprehensive behavior of microgel dispersion is schematically described in Figure 2f.

3.3. Thermogelling Behaviors of p(NiPAm-co-HEMA) Microgel Dispersion

We characterized the thermogelling behaviors of p(NiPAm-co-HEMA) microgel dispersion in accordance with the concentration of NaCl, concentration of microgels and composition of NiPAm and HEMA by the SAOS test results measured with a rotational rheometer (refer to the experimental section for the detailed conditions). Following the conventional method of measuring thermogelling behaviors, the gelation temperature at which the colloidal dispersion changes from sol to gel was determined by observing the cross-over behavior of the storage (G′) and loss (G″) moduli, as shown in Figure 3 [26].

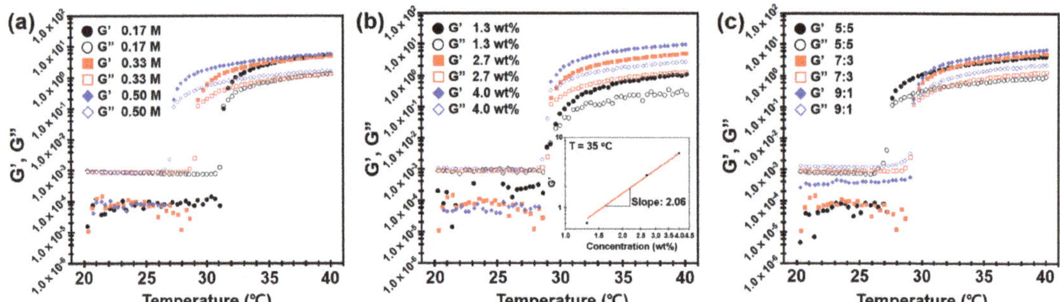

Figure 3. Evolution of storage (G′) and loss (G″) moduli of the p(NiPAm-co-HEMA) microgel dispersion from 20 °C to 40 °C. The procedure was conducted under fixed stress of 0.05 Pa and a frequency of 0.1 Hz ($\omega = 0.63$ rad/s). (**a**) Change in dynamic modulus of the polymer at different salt concentrations. The concentration of polymer and molar ratio between NiPAm and HEMA were kept constant at 2.7 wt% and 7:3, respectively. (**b**) Change in dynamic modulus at different copolymer concentration (c_m). The concentration of NaCl and molar ratio between NiPAm and HEMA were kept constant at 0.17 M and 7:3, respectively. Log plot between G′ and microgel concentration showed a power-law relationship with a slope of 2.06 (inset). (**c**) Change in dynamic modulus at different molar ratios between NiPAm and HEMA. The concentrations of polymer and NaCl were kept constant at 2.7 wt% and 0.33 M, respectively.

3.3.1. Effect of Salt Concentration

From the prior investigation of the dispersion stability, it can be concluded that the microgels under the speculation are stable up to 0.5 M NaCl. Thus, the effect of salt concentration for thermogelling behaviors was investigated for 2.7 wt% of 7:3 molar ratio of p(NiPAm-co-HEMA) microgel dispersion with NaCl concentrations of 0.17 M, 0.33 M, and 0.5 M. As shown in Figure 3a, larger G″ than G′ was observed initially for all the samples with no significant fluctuation, which proves a liquid-like sol state. As temperature kept increasing, a crossover between G′ and G″ occurred at 31.8 °C, 29.3 °C, and 27.3 °C for samples with salt concentrations of 0.17 M, 0.33 M, and 0.5 M, respectively. The result shows an aggregation of p(NiPAm-co-HEMA) microgels and thereby gelation of the system. Based on the findings, it was concluded that the gelation temperature decreases as the salt concentration increases. Decreasing trends of gelation temperature can be attributed to the screening of electrostatic repulsion by the salt addition, which results in the decrease of Debye length of p(NiPAm-co-HEMA) microgels. In addition, it is notable that the plateau of the G′ and G″ after gelation was formed at a similar magnitude for all the samples. It implies that the mechanical strength of gels in different salt concentrations was not affected.

3.3.2. Effect of Microgel Concentration

Effect of microgel concentration was conducted by comparing 1.3 wt%, 2.7 wt%, and 4.0 wt% of 7:3 molar ratio of p(NiPAm-co-HEMA) microgel dispersion at a fixed NaCl concentration of 0.33 M. As shown in Figure 3b, gelation temperature where the crossover between G′ and G″ occurs did not change by the concentration fractions of p(NiPAm-co-HEMA) microgels. On the other hand, the mechanical strength of gels was affected by the concentration. When the temperature exceeds the gelation temperature, the magnitude of both G′ and G″ at plateau were proportional to the microgel concentration. It is noted that the magnitude of G′ and G″ at sol state (i.e., below gelation temperature) only showed little increases with increasing the concentration of p(NiPAm-co-HEMA) microgels. The current results suggest that the viscous properties ($\cong$G″$/\omega$) at the sol states of the samples are all close to that of pure water, i.e., G″~O(10^{-3}) [27], which are expected to be proportional to the volume fractions of the microgel. However, the change in the viscous properties with the increasing microgel concentration is too small to be captured within the sensitivity limit of the rotational rheometer. On the other hand, it is clear that their gel strengths are significantly affected by the formation of denser physical networks between microgels during the gelation process as the microgel concentration increases [26]. The relationship between c_m and gel strength (G′) clearly shows a power-law relationship (inset of Figure 3b), which is consistent with the previous studies on the gelation of particulate systems [28]. The power-law exponent is 1 when the particle volume fraction is close to the gelation particle volume fraction (ϕ_g), which increases to 3–5 as the volume fraction significantly deviates from ϕ_g [28]. Therefore, it can be concluded that the current particle volume fraction ($\approx kc_m$) range is not far from ϕ_g for p(NiPAm-co-HEMA) microgels, in which k is a proportionality constant between the particle concentration and volume fraction.

3.3.3. Effect of NiPAm:HEMA

The effect of NiPAm:HEMA molar ratio in p(NiPAm-co-HEMA) microgel was investigated by preparing 2.7 wt% of p(NiPAm-co-HEMA) microgels that composed of NiPAm and HEMA with the molar ratio of 5:5, 7:3, and 9:1, respectively, at a fixed NaCl concentration of 0.33 M. As shown in Figure 3c, the effect of NiPAm:HEMA molar ratio was somewhat less clear than that of NaCl and microgel concentrations. In both sol and gel states, magnitudes of G′ and G″ were slightly higher with a lower HEMA molar ratio. For gelation temperature, there was a negligible difference between samples with a NiPAm:HEMA molar ratio of 7:3 and 9:1, and a decrease of gelation temperature was observed for the sample with a NiPAm:HEMA molar ratio of 5:5. These results are attributed to the difference in size and thereby the different volume fractions of microgels in the dispersion. As we discussed in Figure 2, the size of microgels was increased due to the good water uptake capability of HEMA. Because every sample was prepared to have the same weight percent of microgels, the volume fraction of microgels will be higher for the larger microgels. In addition, the composition of HEMA can also cause the change in the electric double layer and Debye length of microgels. Considering the aforementioned observations comprehensively, the change in composition affects the gelation properties, but it is not an appropriate variable for precise control of the gelation temperature or mechanical properties of the gels.

3.4. Thermogelling Behaviors of p(NiPAm-co-HEMA) Microgel with Silica Nanoparticle Composites

Investigation of thermogelling behaviors of p(NiPAm-co-HEMA) microgel in the presence of silica nanoparticles was conducted. Here, p(NiPAm-co-HEMA) microgels having a 7:3 molar ratio between NiPAm and HEMA were mixed with LUDOX silica (Aldrich, St. Louis, MO, USA) nanoparticles with weight ratios of 1:1 and 1:5, respectively. Rheological behaviors of samples showed an increasing trend of elastic modulus upon gelation while moderately maintaining volume phase transition temperature. As shown in Figure 4a,b, the microgel–silica nanoparticle composite gel having a 1:1 weight ratio exhibited an increase of G' and G'' by 10 folds in comparison to the gel in the absence of silica nanoparticles. This can be attributed to the jammed silica nanoparticles around the p(NiPAm-co-HEMA) microgels during gelation. As illustrated in Figure 4c, volume phase transition of p(NiPAm-co-HEMA) microgels takes place when the surface moiety of microgels change from hydrophilic to hydrophobic. In this case, the hydrophobic nature of microgels cause them to be percolated, which results in the continuous microgel networks forming the bulk gel. In the presence of silica nanoparticles, it is readily adsorbed on the surface of microgels by electrostatic interaction. As a result, percolated microgel network can be reinforced by jammed silica nanoparticles around the microgels, as illustrated in Figure 4d. To investigate the gelation behaviors further, we observed volume phase transition behaviors of microgels through a transmission optical microscope. A neat p(NiPAm-co-HEMA) microgels at sol state showed homogeneous dispersion, as shown in Figure 4e, and it gradually percolated as the temperature reached 60 °C, shown in Figure 4f (see Video S1 for details). In the case of the microgel/silica nanoparticle composite, however, small aggregates were observed at room temperature (Figure 4g), and they locally aggregated as temperature increased (Figure 4h). These behaviors can be explained by electrostatic attractions among positively charged microgels and negatively charged silica nanoparticles. At the elevated temperature of 60 °C, the global aggregation was observed (Figure 4i), in which larger and locally aggregated colloidal grains were prominent (see Video S2 for details). From the results above, it is concluded that the dispersion stability has to be carefully considered for successful gelation of microgel–nanoparticle composite system. Indeed, we found an unstable sol–gel transition behavior when the concentration of silica nanoparticles was increased further. We conducted experiments for the p(NiPAm-co-HEMA) microgels and silica nanoparticle composites having 1:5 weight ratios. Although global gelation was observed by optical microscope for both samples (Figure S2), we were not able to measure the phase transition behavior by rheometer (Figure S3). This is because too large microgel–silica nanoparticle aggregates hinder homogenous and continuous microgel networks, which implies a delicate control of dispersion stability is crucial for engineering the thermogelling behavior of microgel–nanoparticle composites.

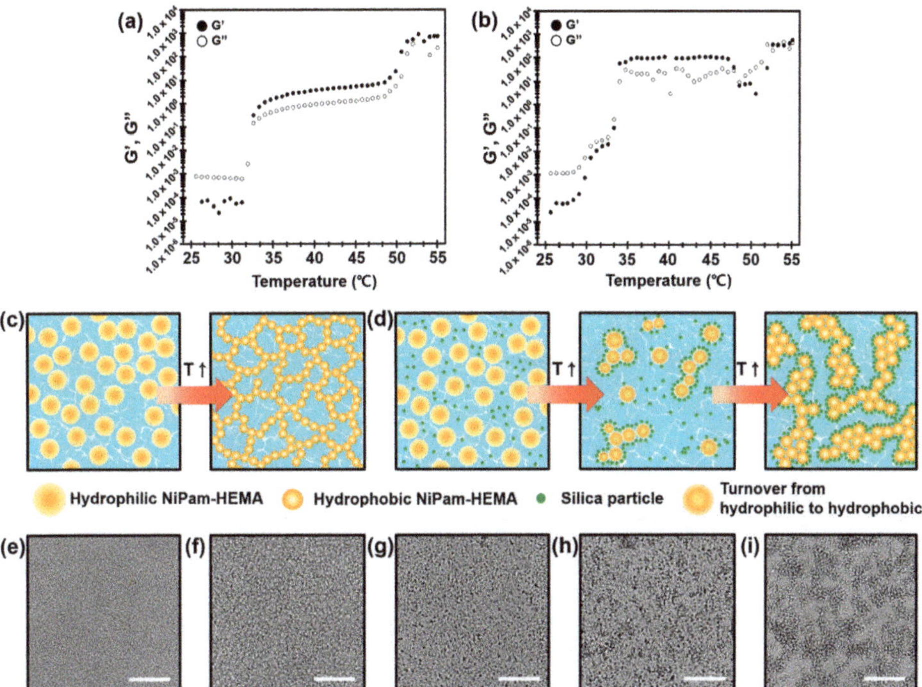

Figure 4. (**a**,**b**) Thermogelling behavior of 2.7 wt% p(NiPAm-co-HEMA) microgels with 7:3 of NiPAm:HEMA molar ratio at 0.17 M of NaCl and silica nanoparticle composite. A comparative study between (**a**) the neat p(NiPAm-co-HEMA) and (**b**) the mixture of the microgel and the silica nanoparticle of 1:1 weight ratio. (**c**,**d**) Schematics of gelation process of (**c**) a neat p(NiPAm-co-HEMA) microgel dispersion and (**d**) the p(NiPAm-co-HEMA) microgel–silica nanoparticle composite. Time-resolved microscopic images showing sol–gel transition of (**e**,**f**) a neat p(NiPAm-co-HEMA) microgels and (**g**–**i**) p(NiPAm-co-HEMA) microgel–silica nanoparticle composite.

4. Conclusions

A reverse sol–gel transition behavior of microgel dispersion has long been studied in the biomedical research field. In particular, a p(NiPAm)-based microgel dispersion was widely studied due to a moderate volume phase transition temperature around human body temperature. In this report, we showed the unique phase transition behavior of p(NiPAm)-based microgel dispersion is maintained in the presence of nanoparticle additives. When the p(NiPAm-co-HEMA) microgel was mixed with silica nanoparticles at a 1:1 weight ratio, it showed a stable phase transition behavior. Although some micro aggregation between microgels and silica nanoparticles was observed, a reversible global phase transition behavior was also observed. It is noted that a bulk gel strength was affected by the existence of additives as the jammed colloidal nanoparticle reinforced the microgel networks. The result implies that the microgel system can be potentially feasible for versatile applications that demand a complex colloidal system. For example, one can introduce the conducting colloids to microgel dispersion to provide self-healing properties to electronic materials. By incorporating plasmonic or fluorescent colloids with microgels, optical signals can be amplified or reduced because the sol–gel transition also drives the volume change of the system. Rheological property can also be tuned by gelation, which can be applied for the formulation of a slurry composed of a complex colloidal system. Therefore, we believe that the microgel–colloid composite system can be a strong candidate to be applied for designing a smart and functional colloidal system in the future.

Supplementary Materials: The following are available online at https://www.mdpi.com/1996-194 4/14/5/1212/s1, Figure S1: (a) UV-vis spectrum of p(NiPAm-co-HEMA) microgel dispersion with molar ratio between NiPAm and HEMA of 7:3 at various pH values. The increase of transmittance at acidic of pH 4 and alkalic of pH 10 is attributed to the sedimentation of microgels. (b,c) photograph images of p(NiPAm-co-HEMA) microgel dispersion at (b) pH 4, and (c) pH 10. Photos were taken after slightly shaking the sedimented microgels solution for the visualization.; Figure S2: Microscopic images of p(NiPAm-co-HEMA) microgel and silica nanoparticle composite. The composite was 1:5 in weight ratio. (a) Small aggregates at room temperature. (b) local aggregation as the temperature increases. (c) Global aggregation and gelation of the composite with 1:5 weight ratio.; Figure S3: Evolution of dynamic modulus of 1:5 weight ratio composite. Turnover of G′(storage modulus) and G″(loss modulus) was observed 51.3 °C.; Video S1: Supporting Movie S1; Video S2: Supporting Movie S2.

Author Contributions: Conceptualization, J.S.K. and T.S.S.; Investigation, J.S.K.; Methodology, B.S.H.; Project administration, T.S.S.; Supervision, J.M.K. and T.S.S.; Writing—original draft, B.S.H. and T.S.S.; Writing—review & editing, J.S.K., J.M.K. and T.S.S. All authors have read and agreed to the published version of the manuscript.

Funding: This study was supported by the Ajou University research fund, grant number [S-2019-G0001-00498].

Institutional Review Board Statement: Not applicable.

Informed Consent Statement: Not applicable.

Data Availability Statement: All data in this study are available from the corresponding author upon reasonable request.

Conflicts of Interest: The authors declare no conflict of interest.

References

1. Tian, H.; Liang, J.; Liu, J. Nanoengineering Carbon Spheres as Nanoreactors for Sustainable Energy Applications. *Adv. Mater.* **2019**, *31*, 1903886. [CrossRef]
2. Chen, K.; Xue, D. Colloidal Supercapattery: Redox Ions in Electrode and Electrolyte. *Chem. Rec. N. Y.* **2017**, *18*, 282–292. [CrossRef] [PubMed]
3. Moon, H.J.; Ko, D.Y.; Park, M.H.; Joo, M.K.; Jeong, B. Temperature-Responsive Compounds as in Situ Gelling Biomedical Materials. *Chem. Soc. Rev.* **2012**, *41*, 4860–4883. [CrossRef]
4. Sarwan, T.; Kumar, P.; Choonara, Y.E.; Pillay, V. Hybrid Thermo-Responsive Polymer Systems and Their Biomedical Applications. *Front. Mater.* **2020**, *7*, 73. [CrossRef]
5. Silan, C.; Akcali, A.; Otkun, M.T.; Ozbey, N.; Butun, S.; Ozay, O.; Sahiner, N. Novel Hydrogel Particles and Their IPN Films as Drug Delivery Systems with Antibacterial Properties. *Colloids Surf. B Biointerfaces* **2012**, *89*, 248–253. [CrossRef]
6. Shim, T.S.; Kim, S.; Heo, C.; Jeon, H.C.; Yang, S. Controlled Origami Folding of Hydrogel Bilayers with Sustained Reversibility for Robust Microcarriers. *Angew. Chem. Int. Ed.* **2012**, *51*, 1420–1423. [CrossRef] [PubMed]
7. Wu, Z.L.; Moshe, M.; Greener, J.; Therien-Aubin, H.; Nie, Z.; Sharon, E.; Kumacheva, E. Three-Dimensional Shape Transformations of Hydrogel Sheets Induced by Small-Scale Modulation of Internal Stresses. *Nat. Commun.* **2013**, *4*, 1586. [CrossRef] [PubMed]
8. Lee, S.; Woods, C.N.; Ibrahim, O.; Kim, S.W.; Pyun, S.B.; Cho, E.C.; Fakhraai, Z.; Park, S.J. Distinct Optical Magnetism in Gold and Silver Probed by Dynamic Metamolecules. *J. Phys. Chem. C* **2020**, *124*, 20436–20444. [CrossRef]
9. Zhang, Y.-Z.; Lee, K.H.; Anjum, D.H.; Sougrat, R.; Jiang, Q.; Kim, H.; Alshareef, H.N. MXenes Stretch Hydrogel Sensor Performance to New Limits. *Sci. Adv.* **2018**, *4*, eaat0098. [CrossRef]
10. Sarwar, M.S.; Dobashi, Y.; Preston, C.; Wyss, J.K.M.; Mirabbasi, S.; Madden, J.D.W. Bend, Stretch, and Touch: Locating a Finger on an Actively Deformed Transparent Sensor Array. *Sci. Adv.* **2017**, *3*, e1602200. [CrossRef]
11. Tang, F.; Ma, N.; Tong, L.; He, F.; Li, L. Control of Metal-Enhanced Fluorescence with PH- and Thermoresponsive Hybrid Microgels. *Langmuir* **2011**, *28*, 883–888. [CrossRef]
12. Farooqi, Z.H.; Naseem, K.; Ijaz, A.; Begum, R. Engineering of Silver Nanoparticle Fabricated Poly (N-Isopropylacrylamide-Co-Acrylic Acid) Microgels for Rapid Catalytic Reduction of Nitrobenzene. *J. Polym. Eng.* **2016**, *36*, 87–96. [CrossRef]
13. Nakao, T.; Nagao, D.; Ishii, H.; Konno, M. Synthesis of Monodisperse Composite Poly(N-Isopropylacrylamide) Microgels Incorporating Dispersive Pt Nanoparticles with High Contents. *Colloids Surf. Physicochem. Eng. Asp.* **2014**, *446*, 134–138. [CrossRef]
14. Sun, J.-Y.; Zhao, X.; Illeperuma, W.R.K.; Chaudhuri, O.; Oh, K.H.; Mooney, D.J.; Vlassak, J.J.; Suo, Z. Highly Stretchable and Tough Hydrogels. *Nature* **2012**, *489*, 133–136. [CrossRef] [PubMed]

15. Timothy, B.; Kim, D.; Yoo, S.I.; Yoon, J. Tuning of Volume Phase Transition for Poly(N-Isopropylacrylamide) Ionogels by Copolymerization with Solvatophilic Monomers. *Soft Matter* **2018**, *14*, 7664–7670. [CrossRef] [PubMed]
16. Wu, J.; Zhou, B.; Hu, Z. Phase Behavior of Thermally Responsive Microgel Colloids. *Phys. Rev. Lett.* **2003**, *90*, 048304. [CrossRef]
17. Zhang, B.; Sun, S.; Wu, P. Synthesis and Unusual Volume Phase Transition Behavior of Poly(N-Isopropylacrylamide)–Poly (2-Hydroxyethyl Methacrylate) Interpenetrating Polymer Network Microgel. *Soft Matter* **2013**, *9*, 1678–1684. [CrossRef]
18. Wu, C.; Zhou, S. Laser Light Scattering Study of the Phase Transition of Poly(N-Isopropylacrylamide) in Water. 1. Single Chain. *Macromolecules* **1995**, *28*, 8381–8387. [CrossRef]
19. Liao, W.; Zhang, Y.; Guan, Y.; Zhu, X.X. Gelation Kinetics of Thermosensitive PNIPAM Microgel Dispersions. *Macromol. Chem. Physic* **2011**, *212*, 2052–2060. [CrossRef]
20. Liu, W.; Zhang, B.; Lu, W.W.; Li, X.; Zhu, D.; Yao, K.D.; Wang, Q.; Zhao, C.; Wang, C. A Rapid Temperature-Responsive Sol–Gel Reversible Poly(N-Isopropylacrylamide)-g-Methylcellulose Copolymer Hydrogel. *Biomaterials* **2004**, *25*, 3005–3012. [CrossRef]
21. Zhang, H.; Sun, L.; Yang, B.; Zhang, Y.; Zhu, S. A Thermo-Responsive Dual-Crosslinked Hydrogel with Ultrahigh Mechanical Strength. *Rsc. Adv.* **2016**, *6*, 63848–63854. [CrossRef]
22. Zhang, B.; Tang, H.; Wu, P. The Unusual Volume Phase Transition Behavior of the Poly(N -Isopropylacrylamide)–Poly (2-Hydroxyethyl Methacrylate) Interpenetrating Polymer Network Microgel: Different Roles in Different Stages. *Polym. Chem.* **2014**, *5*, 5967–5977. [CrossRef]
23. Gan, T.; Zhang, Y.; Guan, Y. In Situ Gelation of P(NIPAM-HEMA) Microgel Dispersion and Its Applications as Injectable 3D Cell Scaffold. *Biomacromolecules* **2009**, *10*, 1410–1415. [CrossRef]
24. Su, G.; Zhou, T.; Liu, X.; Ma, Y. Micro-Dynamics Mechanism of the Phase Transition Behavior of Poly(N-Isopropylacrylamide-Co-2-Hydroxyethyl Methacrylate) Hydrogels Revealed by Two-Dimensional Correlation Spectroscopy. *Polym. Chem.* **2017**, *8*, 865–878. [CrossRef]
25. Yildiz, B.; Işik, B.; Kiş, M. Thermoresponsive Poly(N-Isopropylacrylamide-Co-Acrylamide-Co-2-Hydroxyethyl Methacrylate) Hydrogels. *React. Funct. Polym.* **2002**, *52*, 3–10. [CrossRef]
26. Tung, C.M.; Dynes, P.J. Relationship between Viscoelastic Properties and Gelation in Thermosetting Systems. *J. Appl. Polym. Sci.* **1982**, *27*, 569–574. [CrossRef]
27. Bird, R.B.; Armstrong, R.C.; Hassager, O.; Curtiss, C.F.; Middleman, S. Dynamics of Polymeric Liquids, Vols. 1 and 2. *Phys. Today* **1978**, *31*, 54–57. [CrossRef]
28. Mewis, J.; Wagner, N.J. *Colloidal Suspension Rheology*; Cambridge University Press: Cambridge, UK, 2011. [CrossRef]

Article

Cesium Doping for Performance Improvement of Lead(II)-Acetate-Based Perovskite Solar Cells

Min-Seok Han [1], Zhihai Liu [2], Xuewen Liu [1], Jinho Yoon [1] and Eun-Cheol Lee [1,3,*]

[1] Department of Nano Science and Technology, Graduate School, Gachon University, Gyeonggi 13120, Korea; hanminsuk7@naver.com (M.-S.H.); arelaII5960@gmail.com (X.L.); wlsgh9838@naver.com (J.Y.)

[2] School of Opto-Electronic Information Science and Technology, Yantai University, Yantai 264005, China; zhliu@ytu.edu.cn

[3] Department of Physics, Gachon University, Gyeonggi 13120, Korea

* Correspondence: eclee@gachon.ac.kr; Tel.: +82-31-750-8752; Fax: +82-31-750-8769

Abstract: Lead(II)-acetate ($Pb(Ac)_2$) is a promising lead source for the preparation of organolead trihalide perovskite materials, which avoids the use of inconvenient anti-solvent treatment. In this study, we investigated the effect of cesium doping on the performance of $Pb(Ac)_2$-based perovskite solar cells (PSCs). We demonstrate that the quality of the $CH_3NH_3PbI_3$ perovskite film was improved with increased crystallinity and reduced pinholes by doping the perovskite with 5 mol% cesium. As a result, the power conversion efficiency (PCE) of the PSCs was improved from 14.1% to 15.57% (on average), which was mainly induced by the significant enhancements in short-circuit current density and fill factor. A PCE of 18.02% was achieved for the champion device of cesium-doped $Pb(Ac)_2$-based PSCs with negligible hysteresis and a stable output. Our results indicate that cesium doping is an effective approach for improving the performance of $Pb(Ac)_2$-based PSCs.

Keywords: perovskite solar cells; performance improvement; lead acetate; cesium doping

Citation: Han, M.-S.; Liu, Z.; Liu, X.; Yoon, J.; Lee, E.-C. Cesium Doping for Performance Improvement of Lead(II)-Acetate-Based Perovskite Solar Cells. *Materials* **2021**, *14*, 363. https://doi.org/10.3390/ma14020363

Received: 12 December 2020
Accepted: 11 January 2021
Published: 13 January 2021

Publisher's Note: MDPI stays neutral with regard to jurisdictional claims in published maps and institutional affiliations.

1. Introduction

Organometallic halide perovskites ($APbX_3$, in which A = methylammonium (MA^+) or formamidinium (FA^+), and X = Cl^-, Br^-, or I^-) have attracted considerable attention because of their tunable bandgap, high light absorption coefficient, and long exciton diffusion length over one micrometer [1]. Since being first reported by Kojima et al. in 2009 [2], perovskite solar cells (PSCs) have been intensively investigated, with rapid improvement in power conversion efficiency (PCE) to above 25.5% [3–8]. As a result, PSCs are considered one of the most promising candidates for next-generation solar energy devices.

To prepare a $CH_3NH_3PbI_3$ perovskite, typically lead iodide (PbI_2) is used as the lead source, which chemically reacts with methylammonium iodide (MAI) at a molar ratio of 1:1 [9]. However, in order to obtain a high-quality perovskite film with a uniform morphology, an anti-solvent treatment is needed for the one-step spin-coating process [10]. This anti-solvent treatment requires expensive technology for the stable crystallization of the perovskite grains, which is detrimental to commercialization [11]. To overcome this problem, lead chloride ($PbCl_2$) can be employed to replace PbI_2 as the lead source, which has resulted in PSCs with similar performance to those prepared from PbI_2 [12]. However, to fully remove the residual MACl from the perovskite film, a lengthy thermal annealing process (of about two hours) is required, which consumes a large amount of energy [13]. Besides lead halides, lead acetate ($Pb(Ac)_2$) is another important lead source, which can avoid the need for the inconvenient anti-solvent treatment and lengthy thermal annealing processes [13]. Zhang and coworkers have shown that the $Pb(Ac)_2$-processed $CH_3NH_3PbI_3$ perovskite shows a more uniform and compact morphology with increased crystallinity, compared with $PbCl_2$- or PbI_2-processed perovskites. As a result, the PSCs based on $Pb(Ac)_2$-processed perovskite films were shown to exhibit a high PCE of 14.7%, which is higher than that of either $PbCl_2$- or PbI_2-based PSCs [13]. To improve the performance of PSCs, morphology control of the perovskite film is crucial, because it is strongly related to the charge generation and dissociation properties of the PSCs [14]. Solvent engineering is a widely used technique for controlling the morphology of perovskite films. For example, the use of additional dimethyl-sulfoxide has resulted in an improved film morphology of $Pb(Ac)_2$-based perovskite [15]. Doping of the perovskite crystal is another effective way to improve the morphology of the perovskite films. For example, Br^- and FA^+ have been used to partially replace I^- and MA^+ in the perovskite structure for preparing mix-cation perovskite ($FA_xMA_{1-x}PbI_yBr_{3-y}$), which resulted in a significantly improved PCE and enhanced stability of the PSCs [16]. For PbI_2-processed perovskite, cesium doping into the MA site has been demonstrated to be an efficient way to improve the performance of the associated PSCs [16]. M. Saliba and coworkers have demonstrated that doping with 5 mol% cesium resulted in a uniform and compact perovskite film morphology with fewer pinholes, which significantly improved the PCE of the PSCs from 16.37 to 19.20% [16]. However, there has not been any investigation into the cesium doping effect on the film morphology of $Pb(Ac)_2$-based perovskite and the performance of $Pb(Ac)_2$-based PSCs. Considering the importance and convenience of using $Pb(Ac)_2$ for perovskite preparation, it is crucial to dope $Pb(Ac)_2$-based perovskite with cesium for PSC fabrication.

In this work, we doped the $Pb(Ac)_2$-based perovskite with cesium by adding a small amount of cesium iodide (CsI) into the perovskite precursor. After doping, the perovskite film showed a uniform morphology with enhanced crystallinity and reduced pinholes, which is beneficial for charge transportation. Consequently, the champion device PCE raised from 15.22 to 18.02% with negligible hysteresis and a stable output which is a significant improvement in open-circuit voltage (V_{oc}), short-circuit current density (J_{sc}), and fill factor (FF) via cesium doping. Additionally, the average PCE of the $Pb(Ac)_2$-based PSCs was significantly improved from 14.1 to 15.57%. Our results demonstrate the superior effect of cesium doping on the performance improvement of $Pb(Ac)_2$-based PSCs.

2. Materials and Methods

2.1. Materials

Poly(3,4-ethylenedioxythiophene) polystyrene sulfonate (PEDOT:PSS, P AI 4083) was bought from Heraeus Co. (Hanau, Germany). 6,6-phenyl C61-butyric acid methyl ester (PCBM) was purchased from Nano-C Inc. (Westwood, MA, USA). 2,9-dimethyl-4,7-diphenyl-1,10-phenanthroline (BCP) was obtained from Xi'an Polymer Light Technology Corp (Xi'an, China). CsI was purchased from Sigma Aldrich (St. Louis, MO, USA). Methylammonium Iodide (MAI) was provided by Great Cell Solar (Queanbeyan, Australia). $Pb(Ac)_2$ was provided by Tokyo Chemical Industry (Tokyo, Japan).

2.2. Device Fabrication

Firstly, we cleaned patterned indium tin oxide (ITO)-glass substrates sequentially in detergent, acetone, and 2-propanol for 15 min. The hole transport material, PEDOT:PSS, was deposited on the ITO-glass substrates through a spin-coating process and then annealed at 140 °C for 15 min in air. The perovskite precursors were made by mixing $Pb(Ac)_2$ and MAI in 1 M dimethylformamide. Then, the perovskite precursor with x mol% cesium, where $x = 0, 2.5, 5.0, 7.5$, or 10.0, was spin-coated on the ITO/PEDOT:PSS substrates at a rotation speed of 4000 rpm, which was followed by annealing at 80 °C for 10 min. The electron transport layer (ETL) was formed by spin-coating of $PC_{61}BM$ (30 mg·ml^{-1} in Chlorobenzene) on the perovskite layer at 2000 rpm for 30 s. For better charge transport, we deposited a BCP layer onto the ETL by spin-coating BCP solution (0.5 mg·ml^{-1} in 2-propanol) at 450 rpm for 30 s. Finally, the deposition of an 80 nm silver electrode was achieved with thermal evaporation under a high vacuum of approximately 10^{-6} Torr. The device area was determined by the overlapped rectangle between the ITO and Ag electrode bars, being 0.06 cm^2 (0.2 cm × 0.3 cm).

2.3. Measurements

X-ray diffraction (XRD) and X-ray photoelectron spectroscopy (XPS) were conducted for the perovskite samples using the D8 Advance X-ray diffractometer (Bruker, Billerica, MA, USA) and the K-Alpha X-ray photoelectron spectrometer (Thermo Electron, Waltham, MA, USA), respectively.

Current density–voltage (J–V) curves of the PSCs were obtained using the 2400 Series *J–V* Source Meter (Keithley Instrument, Solon, OH, USA) under an irradiation intensity of 100 mW cm^2 (AM1.5). We used a solar simulator (XES-301S, SAN-EI ELECTRIC, Osaka, Japan) for simulating sunlight irradiation.

The space charge limited current (SCLC) of a hole-only device (glass/ITO/PTAA/ Perovskite/PTAA/Ag) was obtained using the Keithley 2400 Source Meter under dark conditions. Electrochemical impedance spectroscopy (EIS) of the PSCs was performed with an electrochemical work station (CH instruments, Austin, TX, USA) under dark conditions. Steady-state photoluminescence (PL) spectroscopy was conducted using FLS920 (Edinburgh Instruments, Livingston, UK) at wavelengths between 720 nm and 800 nm with the excitation wavelength of 514 nm. Ultraviolet–visible absorption spectroscopy was performed with a UV–vis-NIR 3600 spectrometer (Shimadzu, Kyoto, Japan). The morphology of the devices was measured by the scanning electron microscope (SEM, JOEL, Tokyo, Japan) and atomic force microscope (AFM, Veeco, Plainview, NY, USA).

3. Results

Figure 1a shows the schematic of the PSCs with a standard inverted structure of ITO/PEDOT:PSS/Perovskite/PCBM/BCP/Ag. The Pb(Ac)$_2$-processed perovskite films (with and without cesium doping) were sandwiched between a poly(3,4- ethylenedioxythio- phene) polystyrene sulfonate (PEDOT:PSS) hole transport layer and a 6,6-phenyl-C61- butyric acid methyl ester (PC$_{61}$BM) electron transport layer. The J–V curves of the PSCs with 0–7.5 mol% cesium doping are exhibited in Figure 1b, with the photovoltaic parameters listed in Table 1. The reference PSCs without cesium doping had an average PCE of 14.1%, which is a standard value for Pb(Ac)$_2$-processed inverted PSCs. When the perovskite was doped with 2.5 mol% cesium, the PCE increased to 15.04%. With 5 mol% cesium doping, the PCE further increased to 15.57%, which was mainly induced by the significant improvements in J$_{sc}$ (from 20.16 to 21.08 mA cm^{-2}) and FF (from 0.69 to 0.75). Doping with 7.5 mol% cesium degraded the PCE to 15.37%, indicating that 5% is the optimum cesium doping concentration for maximizing the PCE. As shown in Figure S1, we measured the J–V curves scanned in the reverse and forward directions at a scan rate of 200 mV s^{-1}. The J–V curve of the reverse scan was almost the same as that of forward scan, indicating a negligible hysteresis of the device. To investigate the hysteresis deeply, dynamic J–V scans with calculation of the hysteresis index [17] are required, which is beyond the scope of this study.

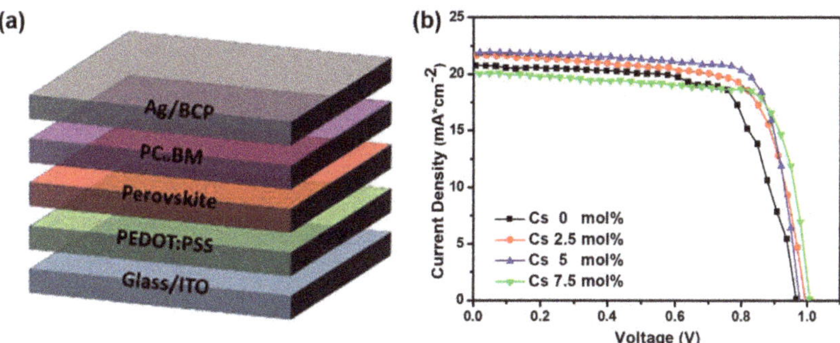

Figure 1. (**a**) Perovskite solar cell structure. (**b**) Current density–voltage (J–V) characteristics of the MAPbI3 PSCs with different cesium doping concentration.

Table 1. Average photovoltaic parameters of the MAPbI$_3$ PSCs based on perovskite precursors with 0 mol%, 2.5 mol%, 5.0 mol%, and 7.5 mol% cesium doping.

Cesium Doping Concentration (mol%)	V_{oc} (V)	J_{sc} (mA cm^{-2})	FF (%)	PCE (%)
0	0.97	20.16	69	14.10
2.5	0.98	21.84	71	15.04
5	0.98	21.08	75	15.57
7.5	1.01	20.8	75	15.37

To identify the origin of the improved PCE by cesium doping, we investigated the morphology of the 5%-Cs-doped and undoped perovskite films. As is evident in the surface SEM images in Figure 2a, the undoped perovskite film had poor surface coverage with many pinholes. A perovskite layer processed from a Pb(Ac)$_2$-based precursor showed a similar surface morphology with some flaws, which might be caused by MA and halide deficiencies as shown in [13]. Figure 2a,d show that with increasing cesium doping concentration, the coverage of the perovskite layer onto PEDOT:PSS increased. As shown in Figure 2c, the 5 mol%-cesium-doped perovskite film had a dense and uniform morphology with full surface coverage. AFM images, shown in Figure S2, further confirm the increased surface uniformity with cesium doping; the root mean square roughness of the cesium-doped perovskite film is 8.6 nm, which is much lower than that of the pristine perovskite (14.6 nm). Furthermore, we found that the cesium-doped perovskite film showed less lateral grain boundaries compared to the pristine perovskite film. As discussed in previous studies, the pinholes in the perovskite film trap carriers, which further increase the charge recombination in the PSCs [18]. The SEM and AFM measurements indicate an improved morphology of the perovskite film upon cesium doping, which explains the PCE improvement, where improved perovskite seeding may be induced by the cesium addition [16]. These seeds might later turn into nucleation sites for further growth of perovskite during crystallization, which results in denser grains [16]. A similar process was found by Li et al. where MAI-modified PbS nanoparticles behaved as growth seeds for highly compact perovskite films [19]. To prove this mechanism, we characterized the crystallinity of the pristine and cesium-doped perovskite films.

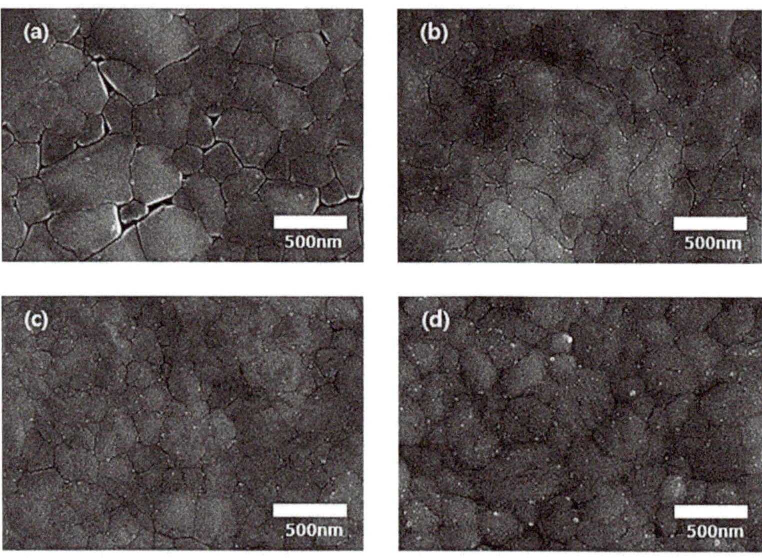

Figure 2. Top-view SEM images of perovskite films with (**a**) 0, (**b**) 2.5, (**c**) 5, and (**d**) 7.5 mol% cesium doping concentration.

Figure 3a compares the XRD spectra of the cesium-doped and pristine perovskite films. All peaks in the XRD patterns show the presence of the CH3NH3PbI3 tetragonal crystal structure [20]. It can be seen that the intensity of the (110) peak at 14° of the cesium-doped perovskite film is higher than that of the undoped one. Moreover, the peak at about 12°, which relates to the (001) lattice planes of hexagonal PbI_2, is dramatically reduced with cesium doping. This indicates that the decomposition of the perovskite to PbI_2 was suppressed by cesium doping [21,22]. In the UV–vis light absorption (Figure 3b), a small blue shift can be observed with cesium doping, indicating a slightly increased optical bandgap, in good agreement with previous studies [23].

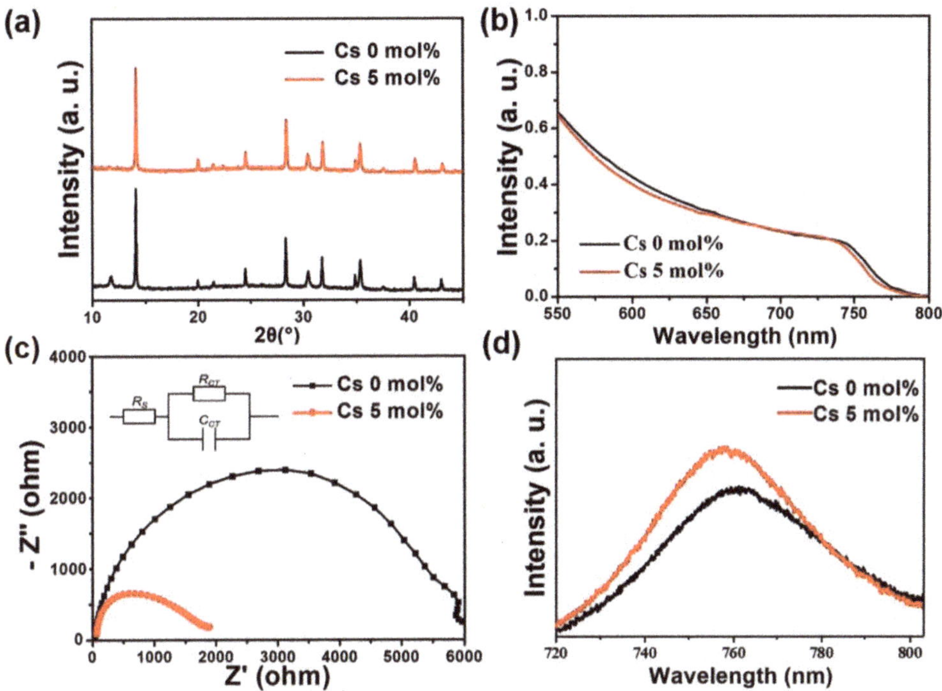

Figure 3. (**a**) XRD patterns. (**b**) Ultraviolet–visible absorption spectra of the perovskite films with different cesium doping concentrations. (**c**) Nyquist plots of the PSCs without and with 5% cesium doping with a bias of 0.8 V. (**d**) Photoluminescence (PL) spectra of the perovskite films without and with 5 mol% cesium doping.

We also conducted EIS for the PSCs under one sun illumination to obtain the resistance information upon cesium doping. Figure 3c shows the Nyquist plots that are fitted with the equivalent circuit, which is shown in the inset. After fitting, the series resistance (R_s), charge recombination resistance (R_{ct}), and chemical capacitance (C_{ct}) of the films could be obtained and the values of them are listed in Table S1. The R_s value for the case with 5 mol% Cs (60.5 Ω) is 33.0% lower than that without Cs doping (90.3 Ω), which contributes to the enhancement of J_{sc} and FF. The C_{ct} values, which are associated with the densities of space charges at the interfaces, are similar for the cases with and without 5% Cs doping (2.9×10^{-9} F and 3.0×10^{-9} F, respectively). The R_{ct} of 5% Cs-doped sample (3198 Ω) is lower than that of undoped sample (6653 Ω). Because the lower R_{ct} indicates the larger electron recombination at the interfaces, the R_{ct} values predict the higher leakage current and the lower J_{sc} for the Cs-doped samples. However, our experimental results show that the doping of 5% Cs reduces the leakage current, as explained below, and increases the J_{sc} (see Figure 4b). The experimental results of previous studies are also controversial; some studies reported that J_{sc} increases with R_{ct} [24–26], while other studies reported the increase in R_{ct} reduced J_{sc} [27–29]. We speculate that R_{ct} in our circuit model may not correctly represent the recombination resistance; R_{ct} in the circuit model was extracted from a high frequency impedance semicircle, whereas some previous studies insisted that R_{ct} is related to both high and low frequency semicircles [30–32]. Further studies using more sophisticated circuit models are required to obtain the more accurate R_{ct}. Figure 3d shows the PL spectra for the 5 mol%-Cs-doped and undoped perovskite films on glass substrates. Evidently, the PL peak of the cesium-doped perovskite film was slightly blue-shifted to 756 nm (the PL peak of the pristine perovskite film is at 760 nm), which is consistent with the UV–vis absorption results in Figure 3d. The intensity of the PL peak of the cesium-doped perovskite film is 24% higher than that of the pristine perovskite film, which indicates decreased surface-trap states (related to non-radiative PL recombination) and increased perovskite crystallinity (consistent with the SEM results) [33].

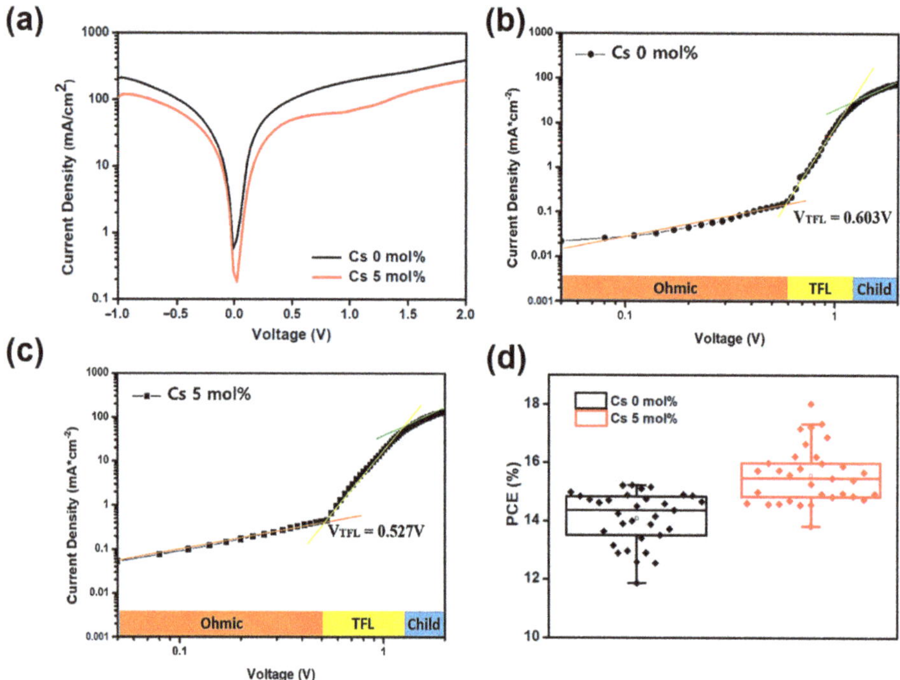

Figure 4. (**a**) Dark J−V characteristics of the PSCs with and without 5 mol% cesium doping. SCLC of the PSCs (**b**) without and (**c**) with cesium doping. (**d**) PCE distribution box chart of the PSCs without and with 5% cesium doping.

Figure 4a showed the dark J–V characteristics of the 5 mol%-Cs-doped and undoped PSCs. The cesium-doped PSC shows smaller leakage current than the reference PSC without cesium doping across the voltage range 0 to −1.0 V. To analyze the trap density of the perovskite films with cesium doping, we measured the SCLC of the hole-only devices described above [34,35]. As shown in Figure 4b,c, the J−V curve can be divided into three regions.

The first segment at low bias (<0.4 V) is the ohmic region, in which the current density shows the almost linear increase with the voltage [36]. The second segment is called the trap-filled limit (TFL) region, in which the current density has rapid nonlinear growth, indicating the TFL in which the injected carriers deactivate available trap states [36]. At high voltages, the current density increases slowly, which is referred to as the Child's regime.

The TFL voltage (V_{TFL}) is the voltage where the ohmic and TFL current curves intersect. The trap density (n_{trap}) can be calculated from V_{TFL} using the following equation [37];

$$V_{TFL} = \frac{en_{trap}L^2}{2\epsilon_0\epsilon} \tag{1}$$

where L is the perovskite film thickness, ϵ ($\approx$5.7565) is the relative dielectric constant of the $CH_3NH_3PbI_3$ perovskite film [38], ϵ_0 is the vacuum permittivity, and e is the elementary charge. As a result, the n_{trap} values of the undoped and 5 mol%-Cs-doped devices are 5.8×10^{16} cm^{-3} and 3.6×10^{16} cm^{-3}, respectively. The reduced trap density in the cesium-doped sample can be explained by the reduced pinholes and improved crystallinity of the perovskite layers (shown in Figure 2). In Figure 4d, it is shown that adding 5 mol% cesium enhanced the average PCE.

XPS spectra for the 5 mol% cesium-doped perovskite film (Figure 5a) show the Cs $3d_{5/2}$ peak at 724.41 eV, confirming the presence of cesium in the sample. In Figure 5b, the cesium doping slightly increases the binding energy of Pb $4f_{5/2}$ from 137.24 to 137.86 eV. For I, the $3d5/2$ peak is also blue-shifted by the cesium doping from 618.16 to 618.8 eV, as shown in Figure 5c. We speculate that the doped cesium atoms cause local distortion in the lattice, which may affect the binding energies of the Pb and I ions.

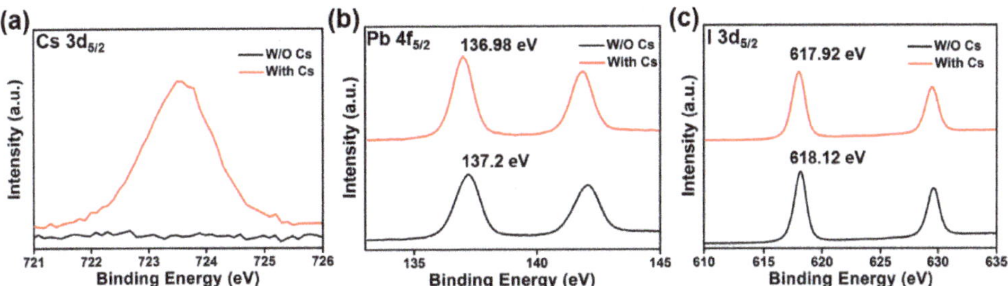

Figure 5. X-ray photoelectron spectroscopy (XPS) results for perovskite films without and with 5% cesium doping: (**a**) Cs $3d_{5/2}$, (**b**) Pb $4f_{5/2}$, and (**c**) I $3d_{5/2}$.

Our results demonstrate that Cs doping is effective for improving the crystallinity and morphology of $Pb(Ac)_2$-based perovskite layers, suppressing the formation of secondary phases such as PbI_2. Thus, Cs doping is promising for enhancing the PCEs of $Pb(Ac)_2$-based PSCs by improving the quality of perovskite films.

4. Summary

In this study, it is proved that suitable amounts of cesium can improve the film morphology and crystallinity of $Pb(Ac)_2$-based perovskite films and adjust the electrical properties of the photoactive layer of perovskite for extracting more charge. PSCs based on these $Pb(Ac)_2$-based perovskite films were demonstrated with a PCE, V_{oc}, J_{sc}, and FF of 15.57%, 0.98 V, 21.08 mA·cm^{-2}, and 0.75, respectively. Additionally, the optimized devices showed negligible hysteresis in the forward and reverse J–V scans. This research shows that a perovskite precursor based on lead acetate is a promising source to achieve highly efficient PSCs, and further improvements will be possible through subtle tuning of the chemical composition.

Supplementary Materials: The following are available online at https://www.mdpi.com/1996-1944/14/2/363/s1, Figure S1: Reverse scan and forward scan J–V curve of 5% Cs-doping device; Figure S2: Tapping-mode AFM height images of (a) the pristine and (b) Cs-doped perovskite films.; Table S1: Fitted values of the equivalent circuit parameters from dark Nyquist plots of devices without and with 5% Cs.

Author Contributions: Conceptualization, Z.L.; Formal analysis, X.L.; Investigation, M.-S.H.; Supervision, E.-C.L.; Visualization, J.Y.; Writing—original draft, M.-S.H., Z.L., X.L. and J.Y.; Writing—review & editing, E.-C.L. All authors have read and agreed to the published version of the manuscript.

Funding: This work was supported by the National Research Foundation of Korea (NRF) funded by the Ministry of Science and ICT (Grant No. NRF-2016R1A2B2015389) and the Gachon University research fund of 2019 (Grant No. GCU-2019-0350).

Data Availability Statement: Data sharing is not applicable to this article.

Conflicts of Interest: The authors declare no conflict of interest.

References

1. Stranks, S.D.; Eperon, G.E.; Grancini, G.; Menelaou, C.; Alcocer, M.J.; Leijtens, T.; Herz, L.M.; Petrozza, A.; Snaith, H.J. Electron-hole diffusion lengths exceeding 1 micrometer in an organometal trihalide perovskite absorber. *Science* **2013**, *342*, 341–344. [CrossRef] [PubMed]
2. Kojima, A.; Teshima, K.; Shirai, Y.; Miyasaka, T. Organometal Halide Perovskites as Visible-Light Sensitizers for Photovoltaic Cells. *J. Am. Chem. Soc.* **2009**, *131*, 6050–6051. [CrossRef] [PubMed]
3. Burschka, J.; Pellet, N.; Moon, S.J.; Humphry-Baker, R.; Gao, P.; Nazeeruddin, M.K.; Gratzel, M. Sequential deposition as a route to high-performance perovskite-sensitized solar cells. *Nature* **2013**, *499*, 316–319. [CrossRef] [PubMed]
4. Zhou, H.; Chen, Q.; Li, G.; Luo, S.; Song, T.B.; Duan, H.S.; Hong, Z.; You, J.; Liu, Y.; Yang, Y. Photovoltaics. Interface engineering of highly efficient perovskite solar cells. *Science* **2014**, *345*, 542–546. [CrossRef] [PubMed]
5. Jeon, N.J.; Noh, J.H.; Yang, W.S.; Kim, Y.C.; Ryu, S.; Seo, J.; Seok, S.I. Compositional engineering of perovskite materials for high-performance solar cells. *Nature* **2015**, *517*, 476–480. [CrossRef] [PubMed]
6. Bi, D.; Yi, C.; Luo, J.; Décoppet, J.-D.; Zhang, F.; Zakeeruddin, S.M.; Li, X.; Hagfeldt, A.; Grätzel, M. Polymer-templated nucleation and crystal growth of perovskite films for solar cells with efficiency greater than 21%. *Nat. Energy.* **2016**, *1*, 16142. [CrossRef]
7. Sahli, F.; Werner, J.; Kamino, B.A.; Brauninger, M.; Monnard, R.; Paviet-Salomon, B.; Barraud, L.; Ding, L.; Diaz Leon, J.J.; Sacchetto, D.; et al. Fully textured monolithic perovskite/silicon tandem solar cells with 25.2% power conversion efficiency. *Nat Mater* **2018**, *17*, 820–826. [CrossRef]
8. NREL. Best Research-Cell Efficiency Chart Ι Photovoltaic Research Ι NREL. 2019. Available online: https://www.nrel.gov/pv/cell-efficiency.html (accessed on 13 January 2021).
9. Kim, H.-S.; Lee, C.-R.; Im, J.-H.; Lee, K.-B.; Moehl, T.; Marchioro, A.; Moon, S.-J.; Humphry-Baker, R.; Yum, J.-H.; Moser, J.E. Lead iodide perovskite sensitized all-solid-state submicron thin film mesoscopic solar cell with efficiency exceeding 9%. *Scientific reports* **2012**, *2*, 1–7. [CrossRef]
10. Paek, S.; Schouwink, P.; Athanasopoulou, E.N.; Cho, K.T.; Grancini, G.; Lee, Y.; Zhang, Y.; Stellacci, F.; Nazeeruddin, M.K.; Gao, P. From nano- to micrometer scale: The role of antisolvent treatment on high performance perovskite solar cells. *Chem. Mater.* **2017**, *29*, 3490–3498. [CrossRef]
11. Tavakoli, M.M.; Yadav, P.; Prochowicz, D.; Sponseller, M.; Osherov, A.; Bulović, V.; Kong, J. Controllable perovskite crystallization via antisolvent technique using chloride additives for highly efficient planar perovskite solar cells. *Adv. Energy Mater.* **2019**, *9*, 1803587. [CrossRef]
12. Pool, V.L.; Gold-Parker, A.; McGehee, M.D.; Toney, M.F. Chlorine in PbCl$_2$-derived hybrid-perovskite solar absorbers. *Chem. Mater.* **2015**, *27*, 7240–7243. [CrossRef]
13. Zhang, W.; Saliba, M.; Moore, D.T.; Pathak, S.K.; Hörantner, M.T.; Stergiopoulos, T.; Stranks, S.D.; Eperon, G.E.; Alexander-Webber, J.A.; Abate, A.; et al. Ultrasmooth organic–inorganic perovskite thin-film formation and crystallization for efficient planar heterojunction solar cells. *Nat. Commun.* **2015**, *6*, 1–10.
14. Zhang, F.; Zhu, K. Additive engineering for efficient and stable perovskite solar cells. *Adv. Energy Mater.* **2020**, *10*, 1902579. [CrossRef]
15. Liu, Y.; Liu, Z.; Lee, E.-C. Dimethyl-sulfoxide-assisted improvement in the crystallization of lead-acetate-based perovskites for high-performance solar cells. *J. Mater. Chem. C* **2018**, *6*, 6705–6713. [CrossRef]
16. Saliba, M.; Matsui, T.; Seo, J.-Y.; Domanski, K.; Correa-Baena, J.-P.; Nazeeruddin, M.K.; Zakeeruddin, S.M.; Tress, W.; Abate, A.; Hagfeldt, A.; et al. Cesium-containing triple cation perovskite solar cells: Improved stability, reproducibility and high efficiency. *Energy Environ. Sci.* **2016**, *9*, 1989–1997. [CrossRef] [PubMed]
17. Nemnes, G.A.; Besleaga, C.; Tomulescu, A.G.; Palici, A.; Pintilie, L.; Manolescu, A.; Pintilie, I. How measurement protocols influence the dynamic JV characteristics of perovskite solar cells: Theory and experiment. *Sol. Energy* **2018**, *173*, 976–983. [CrossRef]
18. Liu, W.; Liu, N.; Ji, S.; Hua, H.; Ma, Y.; Hu, R.; Zhang, J.; Chu, L.; Li, X.; Huang, W. Perfection of perovskite grain boundary passivation by rhodium incorporation for efficient and stable solar cells. *Nano-Micro Lett.* **2020**, *12*, 119. [CrossRef]
19. Li, S.-S.; Chang, C.-H.; Wang, Y.-C.; Lin, C.-W.; Wang, D.-Y.; Lin, J.-C.; Chen, C.-C.; Sheu, H.-S.; Chia, H.-C.; Wu, W.-R.; et al. Intermixing-seeded growth for high-performance planar heterojunction perovskite solar cells assisted by precursor-capped nanoparticles. *Energy Environ. Sci.* **2016**, *9*, 1282–1289. [CrossRef]

20. Sewvandi, G.A.; Hu, D.; Chen, C.; Ma, H.; Kusunose, T.; Tanaka, Y.; Nakanishi, S.; Feng, Q. Antiferroelectric-to-ferroelectric switching in $CH_3NH_3PbI_3$ perovskite and its potential role in effective charge separation in perovskite solar cells. *Phys. Rev. Appl.* **2016**, *6*, 24007. [CrossRef]
21. Gausin, C.M. Improved Thermal Stability of Cesium-Doped Perovskite Films with PMMA for Solar Cell Application. Master's Thesis, Old Dominion University, Norfolk, VA, USA, 2018.
22. Drahansky, M.; Paridah, M.; Moradbak, A.; Mohamed, A.; Owolabi, F.; Asniza, M. We are IntechOpen, the world's leading publisher of Open Access books Built by scientists, for scientists TOP 1%. *Intech* **2016**, *1*, 13.
23. Wu, C.; Guo, D.; Li, P.; Wang, S.; Liu, A.; Wu, F. A study on the effects of mixed organic cations on the structure and properties in lead halide perovskites. *Phys. Chem. Chem. Phys.* **2020**, *22*, 3105–3111. [CrossRef]
24. Yang, D.; Zhou, X.; Yang, R.; Yang, Z.; Yu, W.; Wang, X.; Li, C.; Liu, S.; Chang, R.P.H. Surface optimization to eliminate hysteresis for record efficiency planar perovskite solar cells. *Energy Environ. Sci.* **2016**, *9*, 3071–3078. [CrossRef]
25. Yang, D.; Yang, R.; Zhang, J.; Yang, Z.; Liu, S.; Li, C. High efficiency flexible perovskite solar cells using superior low temperature TiO_2. *Energy Environ. Sci.* **2015**, *8*, 3208–3214. [CrossRef]
26. Wang, S.; Zhu, Y.; Wang, C.; Ma, R. Interface modification by a multifunctional ammonium salt for high performance and stable planar perovskite solar cells. *J. Mater. Chem. A* **2019**, *7*, 11867–11876. [CrossRef]
27. Gao, W.; Ran, C.; Li, J.; Dong, H.; Jiao, B.; Zhang, L.; Lan, X.; Hou, X.; Wu, Z. Robust Stability of Efficient Lead-Free Formamidinium Tin Iodide Perovskite Solar Cells Realized by Structural Regulation. *J. Phys. Chem. Lett.* **2018**, *9*, 6999–7006. [CrossRef]
28. Mahbubur Rahman, M.; Chandra Deb Nath, N.; Lee, J.-J. Electrochemical Impedance Spectroscopic Analysis of Sensitization-Based Solar Cells. *Isr. J. Chem.* **2015**, *55*, 990–1001. [CrossRef]
29. Yu, J.C.; Hong, J.A.; Jung, E.D.; Kim, D.B.; Baek, S.-M.; Lee, S.; Cho, S.; Park, S.S.; Choi, K.J.; Song, M.H. Highly efficient and stable inverted perovskite solar cell employing PEDOT:GO composite layer as a hole transport layer. *Sci. Rep.* **2018**, *8*, 1070. [CrossRef]
30. Pockett, A.; Eperon, G.E.; Sakai, N.; Snaith, H.J.; Peter, L.M.; Cameron, P.J. Microseconds, milliseconds and seconds: Deconvoluting the dynamic behaviour of planar perovskite solar cells. *Phys. Chem. Chem. Phys.* **2017**, *19*, 5959–5970. [CrossRef]
31. Zolfaghari, Z.; Hassanabadi, E.; Pitarch-Tena, D.; Yoon, S.J.; Shariatinia, Z.; van de Lagemaat, J.; Luther, J.M.; Mora-Seró, I. Operation mechanism of perovskite quantum dot solar cells probed by impedance spectroscopy. *ACS Energy Lett.* **2019**, *4*, 251–258. [CrossRef]
32. Zarazua, I.; Sidhik, S.; Lopéz-Luke, T.; Esparza, D.; De la Rosa, E.; Reyes-Gomez, J.; Mora-Sero, I.; Garcia-Belmonte, G. Operating mechanisms of mesoscopic perovskite solar cells through impedance spectroscopy and J–V modeling. *J. Phys. Chem. Lett.* **2017**, *8*, 6073–6079. [CrossRef]
33. Lim, K.-G.; Kim, H.-B.; Jeong, J.; Kim, H.; Kim, J.Y.; Lee, T.-W. Boosting the power conversion efficiency of perovskite solar cells using self-organized polymeric hole extraction layers with high work function. *Adv. Mater.* **2014**, *26*, 6461–6466. [CrossRef]
34. Rose, A. Space-charge-limited currents in solids. *Phys. Rev.* **1955**, *97*, 1538. [CrossRef]
35. Smith, R.W.; Rose, A. Space-charge-limited currents in single crystals of cadmium sulfide. *Phys. Rev.* **1955**, *97*, 1531. [CrossRef]
36. Li, M.; Li, B.; Cao, G.; Tian, J. Monolithic $MAPbI_3$ films for high-efficiency solar cells via coordination and a heat assisted process. *J. Mater. Chem. A* **2017**, *5*, 21313–21319. [CrossRef]
37. Liu, Y.; Yang, Z.; Cui, D.; Ren, X.; Sun, J.; Liu, X.; Zhang, J.; Wei, Q.; Fan, H.; Yu, F.; et al. Two-inch-sized perovskite $CH_3NH_3PbX_3$ (X= Cl, Br, I) crystals: Growth and characterization. *Adv. Mater.* **2015**, *27*, 5176–5183. [CrossRef]
38. Calloni, A.; Abate, A.; Bussetti, G.; Berti, G.; Yivlialin, R.; Ciccacci, F.; Duo, L. Stability of organic cations in solution-processed $CH_3NH_3PbI_3$ perovskites: Formation of modified surface layers. *J. Phys. Chem. C* **2015**, *119*, 21329–21335. [CrossRef]

 *materials*

Article

Fabrication of a Flexible Photodetector Based on a Liquid Eutectic Gallium Indium

Peng Xiao [†], Hyun-Jong Gwak [†] and Soonmin Seo *

Department of Bionano Technology, Gachon University, Seongnam, Gyeonggi 13120, Korea;
zhongpengxiao@gmail.com (P.X.); rhkrguswhd@naver.com (H.-J.G.)
* Correspondence: soonmseo@gachon.ac.kr; Tel.: +82-31-750-8754
† P.X. and H.-J.G. contributed equally to this work.

Received: 27 October 2020; Accepted: 17 November 2020; Published: 18 November 2020

Abstract: A fluidic gallium-based liquid metal (LM) is an interesting material for producing flexible and stretchable electronics. A simple and reliable method developed to facilitate the fabrication of a photodetector based on an LM is presented. A large and thin conductive eutectic gallium indium (EGaIn) film can be fabricated with compressed EGaIn microdroplets. A solution of LM microdroplets can be synthesized by ultrasonication after mixing with EGaIn and ethanol and then dried on a PDMS substrate. In this study, a conductive LM film was obtained after pressing with another substrate. The film was sufficiently conductive and stretchable, and its electrical conductivity was 2.2×10^6 S/m. The thin film was patterned by a fiber laser marker, and the minimum line width of the pattern was approximately 20 μm. Using a sticky PDMS film, a Ga_2O_3 photo-responsive layer was exfoliated from the fabricated LM film. With the patterned LM electrode and the transparent photo-responsive film, a flexible photodetector was fabricated, which yielded photo-response-current ratios of 30.3%, 14.7%, and 16.1% under 254 nm ultraviolet, 365 nm ultraviolet, and visible light, respectively.

Keywords: eutectic gallium indium; EGaIn; liquid metal; gallium alloy; flexible photodetector; flexible electronics

---

1. Introduction

A fluidic gallium-based liquid metal (LM) is an interesting material for flexible and stretchable electronics and has received much attention from researchers owing to its extraordinary electrical conductivity and outstanding mechanical properties [1–4]. It is known that various materials have been utilized for manufacturing flexible and stretchable electronics [5–7]. However, these materials are not flexible and stretchable in the bulk state and need to be treated further. Interestingly, LM has a fluidic nature at room temperature and thus has potential for various applications in stretchable electronics. With the rapid development of artificial and flexible applications and systems, such as flexible and wearable electronics [8,9], electronic skins [10–12], sensors [13,14], and energy harvesting and storage devices [15–17], LMs can be utilized for various applications in these fields. In particular, gallium-based LMs, such as eutectic gallium indium (EGaIn, Ga/In 85.8%/14.2%), have been intensively investigated in recent years because their toxicity is lower than that of mercury. For instance, gallium-based LMs can be used as high-elasticity droplets [18], self-powered liquid metal machines [12], conductive traces for circuit boards [19,20], soft electrodes for plasma [21], and reconfigurable antennas [22,23].

The patterning of LM film is another strategy for the fabrication of wearable, flexible, and stretchable devices. In contrast to other solid metals, manipulation of LM is difficult because of its high surface tension in the fluidic state and quick oxidation in air. To overcome this, various patterning methods for gallium-based LM have been developed. LM electrodes with patterned structures have been developed by many facile and cheap printing methods, including 3D printing [24], direct printing [25], inkjet printing [26], stencil printing [27], photolithography [28], masked deposition [29], microcontact printing [30], laser patterning [31] and dielectrophoresis [32,33]. One of these methods, laser patterning, can be used with various materials and is a fast and simple method for fabricating devices [34,35]. Therefore, we tried to fabricate the desired LM patterns by the laser ablation method. It is expected that a thin LM film can be rapidly patterned by a fiber laser marker without fatal damage to the polydimethylsiloxane (PDMS) substrate because only metals can absorb energy at a wavelength of 1064 nm, while PDMS cannot.

In addition to the fabrication of conductive LM film, another main area of this work is the utilization of a newly formed metal oxide layer of LMs during the process. Most LMs based on gallium alloys are rapidly oxidized in contact with oxygen and form an ultrathin metal oxide layer by a self-limiting reaction [2,4,36]. It is known that a transparent Ga_2O_3 film is used as a photo-responsive film to measure low-density ultraviolet (254 nm and 365 nm) and visible light [37,38]. Furthermore, it has been reported that the oxidized layer could be exfoliated from the LMs with adhesive materials that are used as 2D materials for the semiconducting layer [39]. Thus, it is considered that the newly formed Ga_2O_3 film in this work can be separated with an adhesive material, and this layer would show photo-responsive performance.

In this work, we introduce a simple and reliable method to fabricate a flexible and transparent photodetector based on an LM. A large and thin conductive EGaIn film can be fabricated with compressed EGaIn microdroplets. The LM film is sufficiently conductive and can be rapidly patterned by laser ablation. In addition, a photo-responsive gallium oxide layer can also be separated with an adhesive PDMS substrate from a conductive LM film. A flexible and transparent photodetector was fabricated by combining the patterned LM electrode and the separated Ga_2O_3 film.

2. Materials and Methods

2.1. Fabrication of Liquid Metal (LM) Microdroplets

EGaIn (75.5 wt% Ga and 24.5 wt% In, Sigma-Aldrich, St. Louis, MO, USA) was used to prepare the microdroplets. It consists of Ga and In, and its melting point is 15.7 °C. EGaIn (500 mg) was placed in a 20 mL vial and was filled with ethanol (94.5%, Daejung, Korea). Hereafter, the vial was sonicated using an ultrasonic cleaner (80 W, 40 KHz) for 30 min. After ultrasonication, a suspension of LM microdroplets (<10 μm) was formed, as shown in Figure 1a.

2.2. Preparation of Fully and Incompletely Cured PDMS Substrates

The fully-cured PDMS substrate was prepared by the following process: A PDMS (Dow Corning, Sylgard 184 A/B) mixture with a monomer and a curing agent (at a ratio of 10:1) was prepared and poured onto a flat Petri dish (SPL, Gyeonggi, Korea). The bubbles arising from the vacuum chamber were removed after 1 h, post which it was cured in a convection oven for another 1 h at 80 °C. For an incompletely-cured PDMS substrate, a PDMS mixture with a monomer and a curing agent (at a ratio of 11:1) was used. The mixture was poured on a flat Petri dish, and the height of the incompletely-cured PDMS substrate was 1 mm. It was then cured in an oven for 15 min at 80 °C after removing the bubbles.

2.3. Fabrication of Thin Conductive LM Film with Microdroplet Suspension

A suspension of LM microdroplets formed by ultrasonication was dropped on a flat, fully-cured PDMS substrate and dried at room temperature for 24 h to avoid formation of cracks by rapid solvent evaporation. Subsequently, another flat, fully-cured PDMS substrate was placed onto the dried LM film, which was then pressed by a hydraulic press at 15 MPa for 1 s. After removing the pressure, the upper PDMS substrate was peeled off from the bottom substrate. Finally, thin conductive LM films were left on both the upper and bottom PDMS substrates.

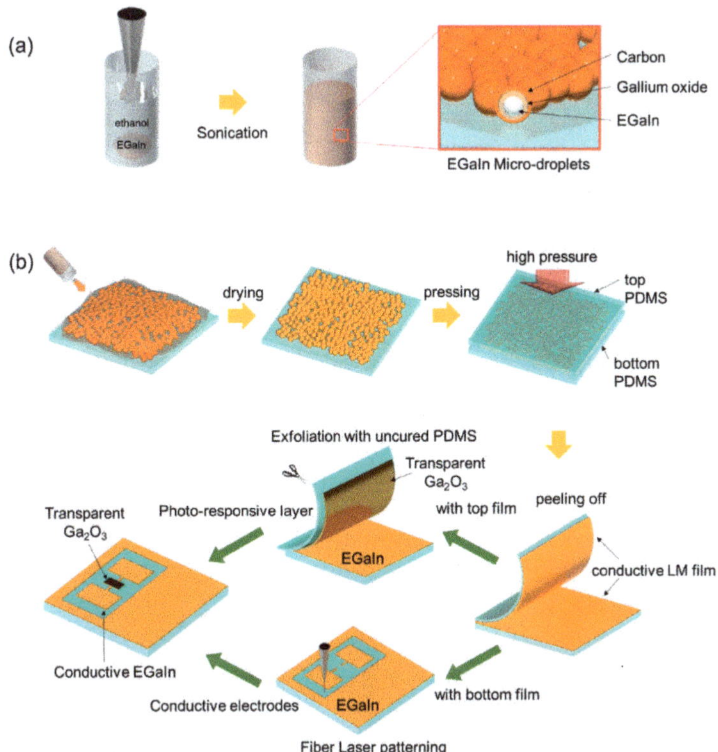

Figure 1. (**a**) Preparation process of EGaIn microdroplets; (**b**) Illustration of the fabrication process of a flexible and transparent photodetector.

2.4. Laser-Engraved Conductive Patterns and Circuits

The thin liquid metal (LM) film was patterned by a fiber laser marker (50 W, Dongil laser technology, Gyeonggi, Korea). The desired circuits and electrodes were fabricated by a subtractive method at a resolution of 20 µm. The scanning speed of the fiber laser marker was 600 mm/s, and the power intensity of the laser was 1% of its maximum power. A pattern with an area of 4 cm² could be engraved within 5 s by a fiber laser marker.

2.5. Fabrication of a Photodetector Based on Oxidized LM Film

An incompletely-cured thin PDMS film was placed slightly on the surface of a gallium-based conductive thin film for conformal contact. The gallium oxide (<10 nm) film on the LM was attached to the incompletely-cured PDMS film, and it was exfoliated from the conductive film after the peeling off process. After cutting down a part of the transparent gallium oxide film on the PDMS film, it was then placed onto the laser-patterned EGaIn electrodes. Finally, the photodetector was fabricated with a transparent gallium oxide film and patterned conductive EGaIn electrodes.

2.6. Characterization

A semiconductor characterization system (Keithley 4200, Beaverton, OR, USA) was used to analyze the electrical properties of the conductive electrodes. Bending and stretching tests were also performed. Photo-detective tests under the irradiation of an ultraviolet (UV) lamp (8 W, Vilber Lourmat, Marne La-Vallee, France) were also carried out at various wavelengths. The tests were also done under a tungsten–halogen lamp (FOK-100 W, Fiber Optic Korea, Cheonan, Korea). Atomic force microscopy (AFM) and scanning electron microscopy (SEM) images were obtained using a multimode AFM (Nanoscope IIIa, Digital Instruments, Bresso, Italy) and FE-SEM (JSM-7500F, Jeol, Tokyo, Japan), respectively.

3. Results

LM based on gallium alloys (EGaIn) was used for the device. In Figure 1a, the preparation process of EGaIn microdroplets is shown. A suspension of LM microdroplets was formed by ultrasonication in ethanol. The inner core of the droplet is EGaIn and is surrounded by gallium oxide with an outer carbon shell [40–42]. In Figure 1b, a schematic illustration of the entire process is shown. The solution was dropped on a flat, fully-cured PDMS substrate and dried at room temperature for 24 h to avoid the formation of cracks by rapid solvent evaporation during the drying process.

LM droplets by ultrasonication were distributed on the PDMS substrate after solvent evaporation, as shown in Figure 2a. EGaIn microdroplets synthesized by ultrasonication were distributed uniformly on the PDMS substrate, and the size of the droplets was smaller than 3 µm, as shown in Figure 2b. Round droplets were observed, and the droplets were wrapped with oxidized gallium material. The film with stacked LM droplets itself is not conductive because it is difficult for an electron to move a long distance through the nonconductive oxidized layers from one droplet to another. The advantage of using the droplets formed by ultrasonication is that it is possible to fabricate a thinner and more uniform LM film when the droplets are used. With bulk EGaIn, it is difficult to create a thin film because the high surface tension of a bulk LM makes it difficult to manage the LM.

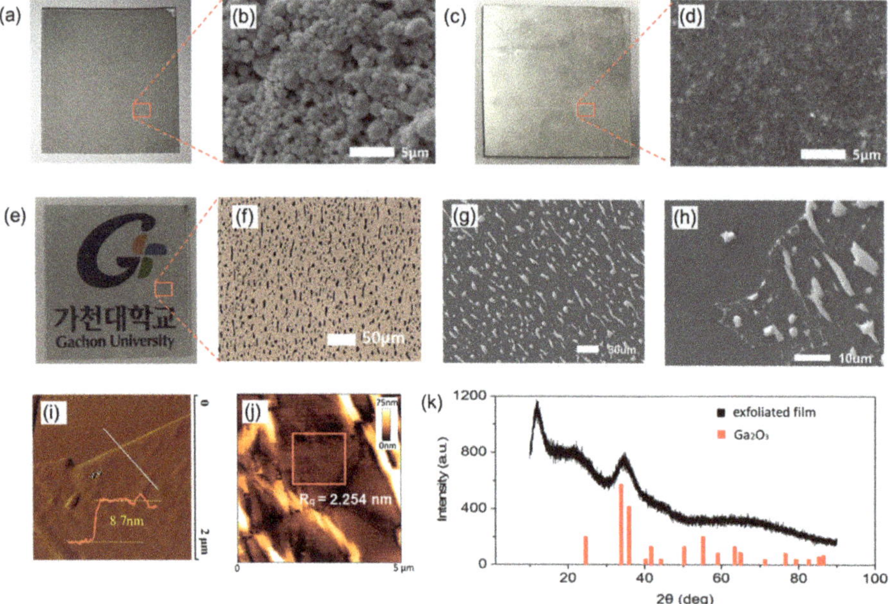

Figure 2. (**a**) Liquid metal microdroplets distributed on a PDMS substrate; (**b**) SEM image of the liquid metal (LM) microdroplets; (**c**) Image of a thin conductive LM film after peeling off process; (**d**) SEM image of the continuous conductive LM film on a PDMS substrate; (**e**) The transparent gallium oxide film attached on a PDMS after peeling off from the LM film; (**f**) Magnified optical image; (**g**) SEM image of the transparent film of (**e**); (**h**) a piece of the thin tore transparent film on a PDMS substrate; (**i**) The thickness and (**j**) the roughness of transparent gallium oxide film (boxed area) by AFM; (**k**) XRD pattern of the exfoliated film on a PDMS substrate.

Subsequently, another flat, fully-cured PDMS substrate was placed onto the dried microdroplet film, which was then pressed by a hydraulic press at a pressure of 15 MPa. The droplets were squashed out and connected to each other by pressure after breaking the oxidized layers of the droplets. The oxidized parts remained in the film inside. However, the connected droplets formed a large conductive thin film, as shown in Figure 2c,d. The amount of oxidized material is much lower than that of the conductive part, and the film is sufficiently conductive for use in electric devices. The top PDMS substrate on the sandwiched LM film was then peeled off. After that, half of the LM remained on the bottom substrate, and another half was transferred onto the top substrate. As a result of the peeling process, two conductive LM sheets were obtained on the top and bottom PDMS substrates. The LM film was formed on the flat PDMS substrate over the entire area. It is difficult to fabricate a very thin LM film from bulk LM because of its high surface tension. However, this difficulty has been overcome by using LM microdroplets. In this work, one of the LM-coated sheets (bottom) was used as a conductive electrode, and another (top) was used for utilizing a photo-responsive layer.

Moreover, an extremely thin oxidized layer could be separated from the LM films by peeling off with a sticky PDMS film. The incompletely-cured PDMS film with strong adhesion was designed to peel off the oxidized layer from the surface of the LM thin film for large-area fabrication. The incompletely-cured PDMS film was placed and covered on the surface of the LM thin film, and it was brought into contact with the oxidized layer of the LM, forming a conformal contact. Hereafter, the transparent oxidized layer was sliced out from the LM film, at the time of peeling off the uncured PDMS film. The optical image of the separated gallium oxide film from the LM film is shown in Figure 2e. The film is very thin and transparent. As shown in Figure 2f,g, small LM spots (<10 μm) remained on the transparent film. However, all the spots were surrounded by gallium oxide film and isolated from each other. Thus, the separated gallium oxide film is laterally nonconductive. The transparent nonconductive film was analyzed by grazing incidence X-ray diffraction (GIXRD). The oxidized sample was prepared on a PDMS substrate, and the graph obtained by GIXRD is shown in Figure 2k. The baseline was similar to reported GIXRD data of bare PDMS [43], and a broad peak was observed around 35°. It is known that GIXRD peaks are also observed at 34° and 36° for a gallium oxide material. Thus, it is concluded that the transparent film is gallium oxide, since the peaks of gallium oxide appear on the graph, and the material was also responsive to UV light in this work. Finally, the separated transparent film was used for fabricating a photodetector because the oxidized thin film is a material mainly based on gallium oxide and it shows high photo-detective property, as reported previously [37,38]. The separated oxidized film on a PDMS substrate could be easily cut using a scissor. Then, it was placed on the substrate between the cathode and anode to fabricate a photodetector. After placing the gallium oxide on the PDMS film between the electrodes, the photodetector with transparent film and flexible LM electrodes was completed, as shown in Figure 1b.

In addition, to fabricate a desirable pattern for a flexible and stretchable device, a fiber laser marker ($\lambda \sim$ 1064 nm) was used for designing the patterns on a conductive LM film. The fabricated conductive films are shown in Figure 3a. A laser engraving method is an efficient way to establish flexible electrodes and patterns for devices. The SEM images in Figure 3b show the pattern with various sizes based on EGaIn by a fiber laser marker. The advantages of using a fiber laser marker for the patterning process are fine pattern resolution and less damage to transparent substrates, such as PDMS or glass, during the patterning process. In fact, buckling of the PDMS substrate due to heat was observed when LM was blazed out. The method can establish a sub-100 μm pattern, and the minimum line was approximately ~20 μm in the experiment. Furthermore, better resolution can be achieved using the laser blazing method if highly qualified equipment is used for patterning [44]. In a previous report, a CO_2 laser, not a fiber laser, was used for cutting the LM electrode inside the PDMS substrate [31]. Actually, the CO_2 laser blazed out the PDMS, and not the LM, in the experiment. In this work, different mechanisms were used. It is known that metal substrates absorb only a small amount of energy from the CO_2 laser, and most of the energy from the CO_2 laser is reflected [45]. In contrast, a metal can absorb the energy of a fiber laser. Thus, the fiber laser is suitable for patterning thin LM films. It could easily remove a thin LM film (<1 μm) quickly. Complex patterns with a resolution below 100 μm (~20 μm) could be made using a fiber laser in this work.

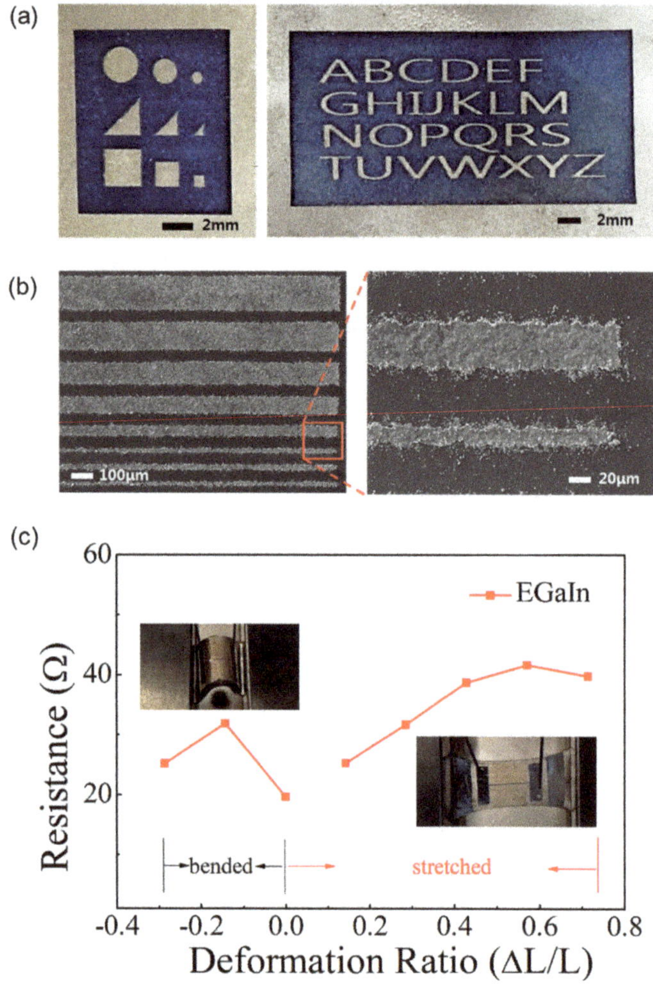

Figure 3. (**a**) Images of a conductive EGaIn film patterned by a fiber laser marker; (**b**) SEM images of conductive EGaIn film patterned by a fiber laser marker, showing a maximum resolution of the pattern of approximately 20 μm; (**c**) Resistance of the EGaIn electrode under bending and stretching.

After peeling off the top PDMS substrate, half of the LM remained on the bottom substrate and another half was transferred to the top substrate, as shown in Figure 1b. As a result of the peeling process, two conductive LM sheets were obtained. Resistance measurement during the stretching test was also performed with an LM electrode (5 mm long and 80 μm wide). It was measured by a semiconductor parameter analyzer, and the characteristic performances of the flexible electrodes are shown in Figure 3c. Resistance of the EGaIn film is between 19.7 Ω and 41.7 Ω; this increases slightly when the film is stretched to 170% of its initial length. The electrical conductivity of the EGaIn film in this work was 2.2×10^6 S/m. The value is approximately two-thirds of its known value (pure EGaIn, ~3.4×10^6 S/m) and was measured using a resistivity meter (Loresta-GX MCP-T700, Mitsubishi Chemical Analytech, Yamato, Japan) with a four-pin probe to overcome the effect of contact resistance. According to the results, the film is sufficiently conductive to be used as an electrode in the circuit. Here, one of the LM sheets was used as a conductive electrode and the other was used for a photo-responsive layer. The thickness of the LM films measured by AFM was approximately 600 nm. It is known that it is difficult to fabricate a very thin LM film because of its high surface tension. However, we could overcome this difficulty by using LM droplets and making a thin (< 1 μm) LM film on the substrate.

4. Discussion

4.1. Fabrication of the Photodetector

In Figure 2b, a conductive LM thin film and a transparent gallium oxide film that is separated from the LM film are shown. All films were fabricated on a large PDMS substrate (5 × 5 cm). As shown in Figure 2d, the morphology of the conductive LM film was not smooth because LM was immediately oxidized and solidified when exposed to air after the peeling-off process. The thickness of the conductive LM film measured by AFM was approximately 600 nm. To measure the thickness of the exfoliated gallium oxide layer, the transparent film on the PDMS substrate was transferred onto a silicon substrate. The measured thickness of the exfoliated metal oxide film by AFM was 8.7 nm, as shown in Figure 2i, and the surface roughness (RMS roughness) of the exfoliated 2D Ga_2O_3 layer for a flat area was 2.254 nm, as shown in Figure 2j. It is assumed that the measured value of the gallium oxide layer is thicker than the known value (~3 nm) because there is further oxidation during the separation process. In this work, the separated transparent film was used as an active layer in a photodetector because the oxidized film based on the gallium oxide is highly photo-detective.

4.2. Device Characterization

A photo-sensitive device was fabricated with flexible electrodes and a transparent gallium oxide film based on LM(EGaIn), as shown in Figure 4a. The black rectangular areas on the right side of the figure are the LM electrodes patterned by a fiber laser, and the transparent area is a gallium oxide film beneath the PDMS substrate. The center image in Figure 4a shows the bird's-eye view of the full structure of the device.

It is known that Ga_2O_3 has a wide bandgap (4.5~4.9 eV) at room temperature [46,47]. As a result, the device is used to measure ultraviolet and visible light as a high-range photodetector. As shown in Figure 4b, the characteristics show an obvious photo-responsive performance under periodic illumination. Photo-responsive tests were also performed with light of three different wavelengths. The devices were illuminated by light periodically at intervals of 30 s with three wavelengths: 254 nm, 365 nm, and visible light. The rise/decay times of the device under illumination of 254 nm, 365 nm, and visible light were 28.2 s/26.7 s, 18.3 s/21.9 s, and 29.1 s/23.6 s, respectively. The responsivities (R) under illumination of 254 nm, 365 nm, and the visible light at 1 V were 2.8×10^{-2} A/W, 3.3×10^{-3} A/W, and 2.6×10^{-6} A/W, respectively. As shown in Figure 4c, the device based on EGaIn shows a photo-response-current ratio ($\Delta I/I_0$) of 30.3% under 254 nm ultraviolet light with an intensity of 0.1 mW/cm^2. It also shows photo-response–current ratios of approximately 14.7% and 16.1% under the illumination of a 365 nm ultraviolet and an ordinary visible light, respectively.

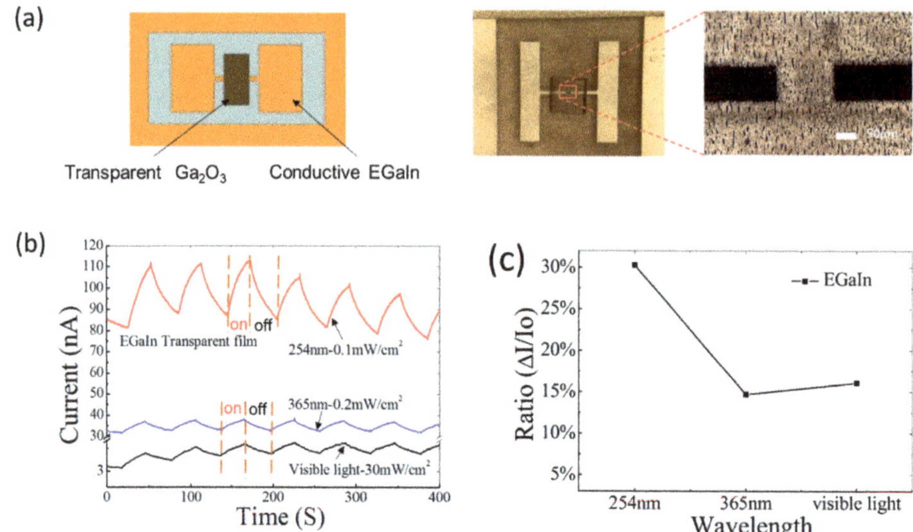

Figure 4. (**a**) Images of the photodetector combined with the conductive electrode and photo-responsive film; (**b**) Time-dependent photo-response curves and (**c**) photo-response-current ratios of the photodetector under illumination with ultraviolet light (λ~254 nm and 365 nm) and visible light. The bias voltage was 0.1 V, and the on/off time of lights was 30 s/30 s.

5. Conclusions

This work describes a new type of flexible photodetector based on a liquid gallium alloy. A simple and reliable method was introduced to fabricate a flexible and transparent photodetector based on LMs. The photodetector was fabricated with a material, EGaIn. Both a conductive electrode and a photo-responsive layer could be obtained from the material and fabricated on PDMS substrates. The fabrication process of a conductive film based on LM microdroplets is an efficient method to fabricate a large-area (5 × 5 cm), flexible, and stretchable LM film. The laser ablation method was also used to fabricate flexible and stretchable electrodes, and the width of the patterned electrodes could be controlled at a level of 20 μm. A photo-responsive layer (~8.7 nm) was exfoliated with an incompletely-cured PDMS by peeling off from the surface of the oxidized LM film. Finally, the photodetector could be made by combining the patterned electrodes and the photo-responsive film. It shows 30.3%, 14.7%, and 16.1% of the photo-response–current ratio under wavelengths of 254 nm and 365 nm in the ultraviolet band, and ordinary visible light, respectively.

The key contribution of this method is that a photosensitive device was fabricated with one material. The semiconducting active layer was exfoliated from conductive materials, and both layers were used in the same device. The laser ablation method shows high performance of controllable, accurate, and efficient patterning. It is expected that the results with LM and various techniques in this work will contribute to advances in the fields of flexible and stretchable sensors.

Author Contributions: Conceptualization, P.X. and S.S.; methodology, P.X. and H.-J.G.; formal analysis, P.X. and H.-J.G., S.S.; investigation, P.X. and H.-J.G.; data curation, P.X. and H.-J.G.; writing—original draft preparation, P.X. and H.-J.G.; writing—review and editing, S.S.; visualization, P.X. and H.-J.G., S.S.; supervision, S.S. All authors have read and agreed to the published version of the manuscript.

Funding: This work was supported by the Basic Science Research Program through the National Research Foundation of Korea (NRF-2018R1D1A1B07041253) and by the Gachon University research fund of 2020(GCU-202004460001).

Conflicts of Interest: The authors declare no conflict of interest.

References

1. Zavabeti, A.; Ou, J.Z.; Carey, B.J.; Syed, N.; Orrell-Trigg, R.; Mayes, E.L.; Xu, C.; Kavehei, O.; O'Mullane, A.P.; Kaner, R.B. A liquid metal reaction environment for the room-temperature synthesis of atomically thin metal oxides. *Science* **2017**, *358*, 332–335. [CrossRef]

2. Kim, Y.D.; Hone, J. Materials science: Screen printing of 2D semiconductors. *Nature* **2017**, *544*, 167–168. [CrossRef] [PubMed]

3. Daeneke, T.; Khoshmanesh, K.; Mahmood, N.; De Castro, I.; Esrafilzadeh, D.; Barrow, S.; Dickey, M.; Kalantar-Zadeh, K. Liquid metals: Fundamentals and applications in chemistry. *Chem. Soc. Rev.* **2018**, *47*, 4073–4111. [CrossRef] [PubMed]

4. Carey, B.J.; Ou, J.Z.; Clark, R.M.; Berean, K.J.; Zavabeti, A.; Chesman, A.S.; Russo, S.P.; Lau, D.W.; Xu, Z.Q.; Bao, Q.; et al. Wafer-scale two-dimensional semiconductors from printed oxide skin of liquid metals. *Nat. Commun.* **2017**, *8*, 1–10. [CrossRef] [PubMed]

5. Shin, J.; Jeong, B.; Kim, J.; Nam, V.B.; Yoon, Y.; Jung, J.; Hong, S.; Lee, H.; Eom, H.; Yeo, J.; et al. Sensitive Wearable Temperature Sensor with Seamless Monolithic Integration. *Adv. Mater.* **2020**, *32*, 1905527. [CrossRef] [PubMed]

6. Kim, J.Y.; Oh, J.Y.; Lee, T.I. Multi-dimensional nanocomposites for stretchable thermoelectric applications. *Appl. Phys. Lett.* **2019**, *114*, 043902. [CrossRef]

7. Suh, Y.D.; Jung, J.; Lee, H.; Yeo, J.; Hong, S.; Lee, P.; Lee, D.; Ko, S.H. Nanowire reinforced nanoparticle nanocomposite for highly flexible transparent electrodes: Borrowing ideas from macrocomposites in steel-wire reinforced concrete. *J. Mater. Chem. C* **2017**, *5*, 791–798. [CrossRef]

8. Hong, Y.J.; Jeong, H.; Cho, K.W.; Lu, N.; Kim, D.H. Wearable and implantable devices for cardiovascular healthcare: From monitoring to therapy based on flexible and stretchable electronics. *Adv. Funct. Mater.* **2019**, *29*, 1808247. [CrossRef]

9. Chen, L.Y.; Tee, B.C.K.; Chortos, A.L.; Schwartz, G.; Tse, V.; Lipomi, D.J.; Wong, H.S.P.; McConnell, M.V.; Bao, Z. Continuous wireless pressure monitoring and mapping with ultra-small passive sensors for health monitoring and critical care. *Nat. Commun.* **2014**, *5*, 1–10. [CrossRef]

10. Wang, S.; Xu, J.; Wang, W.; Wang, G.J.N.; Rastak, R.; Molina-Lopez, F.; Chung, J.W.; Niu, S.; Feig, V.R.; Lopez, J.; et al. Skin electronics from scalable fabrication of an intrinsically stretchable transistor array. *Nature* **2018**, *555*, 83–88. [CrossRef]

11. Yang, J.C.; Mun, J.; Kwon, S.Y.; Park, S.; Bao, Z.; Park, S. Electronic Skin: Recent Progress and Future Prospects for Skin-Attachable Devices for Health Monitoring, Robotics, and Prosthetics. *Adv. Mater.* **2019**, *31*, 1904765. [CrossRef]

12. Yang, Y.; Han, J.; Huang, J.; Sun, J.; Wang, Z.L.; Seo, S.; Sun, Q. Stretchable Energy-Harvesting Tactile Interactive Interface with Liquid-Metal-Nanoparticle-Based Electrodes. *Adv. Funct. Mater.* **2020**, *30*, 1909652. [CrossRef]

13. Kang, H.; Zhao, C.; Huang, J.; Ho, D.H.; Megra, Y.T.; Suk, J.W.; Sun, J.; Wang, Z.L.; Sun, Q.; Cho, J.H. Fingerprint-Inspired Conducting Hierarchical Wrinkles for Energy-Harvesting E-Skin. *Adv. Funct. Mater.* **2019**, *29*, 1903580. [CrossRef]

14. Kim, M.; Seo, S. Flexible pressure and touch sensor with liquid metal droplet based on gallium alloys. *Mol. Cryst. Liq.* **2019**, *685*, 40–46. [CrossRef]

15. Wu, C.; Wang, A.C.; Ding, W.; Guo, H.; Wang, Z.L. Triboelectric nanogenerator: A foundation of the energy for the new era. *Adv. Energy Mater.* **2019**, *9*, 1802906. [CrossRef]

16. Kim, J.H.; Seo, S. Fabrication of an imperceptible liquid metal electrode for triboelectric nanogenerator based on gallium alloys by contact printing. *Appl. Surf. Sci.* **2020**, *509*, 145353. [CrossRef]

17. Park, T.H.; Kim, J.H.; Seo, S. Facile and Rapid Method for Fabricating Liquid Metal Electrodes with Highly Precise Patterns via One-Step Coating. *Adv. Funct. Mater.* **2020**, *30*, 2003694. [CrossRef]

18. Chen, Y.; Liu, Z.; Zhu, D.; Handschuh-Wang, S.; Liang, S.; Yang, J.; Kong, T.; Zhou, X.; Liu, Y.; Zhou, X. Liquid metal droplets with high elasticity, mobility and mechanical robustness. *Mater. Horiz.* **2017**, *4*, 591–597. [CrossRef]

19. Fassler, A.; Majidi, C. Liquid-phase metal inclusions for a conductive polymer composite. *Adv. Mater.* **2015**, *27*, 1928–1932. [CrossRef]

20. Dickey, M.D. Stretchable and soft electronics using liquid metals. *Adv. Mater.* **2017**, *29*, 1606425. [CrossRef]

21. Yu, Y.; Wang, Q.; Wang, X.; Wu, Y.; Liu, J. Liquid metal soft electrode triggered discharge plasma in aqueous solution. *Rsc Adv.* **2016**, *6*, 114773–114778. [CrossRef]
22. Cheng, S.; Wu, Z. Microfluidic stretchable Rf Electron. *Lab Chip* **2010**, *10*, 3227–3234. [CrossRef]
23. Elassy, K.S.; Akau, T.K.; Shiroma, W.A.; Seo, S.; Ohta, A.T. Low-cost rapid fabrication of conformal liquid-metal patterns. *Appl. Sci.* **2019**, *9*, 1565. [CrossRef]
24. Parekh, D.P.; Ladd, C.; Panich, L.; Moussa, K.; Dickey, M.D. 3D printing of liquid metals as fugitive inks for fabrication of 3D microfluidic channels. *Lab Chip* **2016**, *16*, 1812–1820. [CrossRef]
25. Wang, L.; Liu, J. Pressured liquid metal screen printing for rapid manufacture of high resolution electronic patterns. *Rsc Adv.* **2015**, *5*, 57686–57691. [CrossRef]
26. Kim, D.; Yoo, J.H.; Lee, Y.; Choi, W.; Yoo, K.; Lee, J.B. Gallium-based liquid metal inkjet printing. In Proceedings of the IEEE 27th International Conference on Micro Electro Mechanical Systems (MEMS), San Francisco, CA, USA, 26–30 January 2014; pp. 967–970.
27. Lazarus, N.; Bedair, S.S.; Kierzewski, I.M. Ultrafine pitch stencil printing of liquid metal alloys. *Acs Appl. Mater. Interfaces* **2017**, *9*, 1178–1182. [CrossRef]
28. Park, C.W.; Moon, Y.G.; Seong, H.; Jung, S.W.; Oh, J.Y.; Na, B.S.; Park, N.M.; Lee, S.S.; Im, S.G.; Koo, J.B. Photolithography-based patterning of liquid metal interconnects for monolithically integrated stretchable circuits. *Acs Appl. Mater. Interfaces* **2016**, *8*, 15459–15465. [CrossRef]
29. Kramer, R.K.; Majidi, C.; Wood, R.J. Masked deposition of gallium-indium alloys for liquid-embedded elastomer conductors. *Adv. Funct. Mater.* **2013**, *23*, 5292–5296. [CrossRef]
30. Tabatabai, A.; Fassler, A.; Usiak, C.; Majidi, C. Liquid-phase gallium–indium alloy electronics with microcontact printing. *Langmuir* **2013**, *29*, 6194–6200. [CrossRef]
31. Lu, T.; Finkenauer, L.; Wissman, J.; Majidi, C. Rapid prototyping for soft-matter electronics. *Adv. Funct. Mater.* **2014**, *24*, 3351–3356. [CrossRef]
32. Tang, S.Y.; Zhu, J.; Sivan, V.; Gol, B.; Soffe, R.; Zhang, W.; Mitchell, A.; Khoshmanesh, K. Creation of Liquid Metal 3D Microstructures Using Dielectrophoresis. *Adv. Funct. Mater.* **2015**, *25*, 4445–4452. [CrossRef]
33. Krisnadi, F.; Nguyen, L.L.; Ankit, J.M.; Kulkarni, M.R.; Mathews, M.; Dickey, M.D. Directed Assembly of Liquid Metal–Elastomer Conductors for Stretchable and Self-Healing Electronics. *Adv. Mater.* **2020**, *32*, 202001642. [CrossRef] [PubMed]
34. Nam, V.B.; Giang, T.T.; Koo, S.; Rho, J.; Lee, D. Laser digital patterning of conductive electrodes using metal oxide nanomaterials. *Nano Converg.* **2020**, *7*, 1–17. [CrossRef] [PubMed]
35. Nam, V.B.; Shin, J.; Yoon, Y.; Giang, T.T.; Kwon, J.; Suh, Y.D.; Yeo, J.; Hong, S.; Ko, S.H.; Lee, D. Highly Stable Ni-Based Flexible Transparent Conducting Panels Fabricated by Laser Digital Patterning. *Adv. Funct. Mater.* **2019**, *29*, 1806895. [CrossRef]
36. Kim, D.; Thissen, P.; Viner, G.; Lee, D.W.; Choi, W.; Chabal, Y.J.; Lee, J.B. Recovery of nonwetting characteristics by surface modification of gallium-based liquid metal droplets using hydrochloric acid vapor. *Acs Appl. Mater. Interfaces* **2013**, *5*, 179–185. [CrossRef] [PubMed]
37. Hu, G.; Shan, C.; Zhang, N.; Jiang, M.; Wang, S.; Shen, D. High gain Ga_2O_3 solar-blind photodetectors realized via a carrier multiplication process. *Opt. Express* **2015**, *23*, 13554–13561. [CrossRef] [PubMed]
38. Kwon, Y.; Lee, G.; Oh, S.; Kim, J.; Pearton, S.J.; Ren, F. Tuning the thickness of exfoliated quasi-two-dimensional β-Ga_2O_3 flakes by plasma etching. *Appl. Phys. Lett.* **2017**, *110*, 131901. [CrossRef]
39. Hwang, W.S.; Verma, A.; Peelaers, H.; Protasenko, V.; Rouvimov, S.; Xing, H.; Seabaugh, A.; Haensch, W.; Walle, C.V.; Galazka, Z.; et al. High-voltage field effect transistors with wide-bandgap β-Ga_2O_3 nanomembranes. *Appl. Phys. Lett.* **2014**, *104*, 203111. [CrossRef]
40. Lei, N.; Huang, Z.; Rice, S.A.; Grayce, C.J. In-plane structure of the liquid–vapor interfaces of dilute bismuth: Gallium alloys: X-ray-scattering studies. *J. Chem. Phys.* **1996**, *105*, 9615–9624. [CrossRef]
41. Boley, J.W.; White, E.L.; Kramer, R.K. Mechanically sintered gallium–indium nanoparticles. *Adv. Mater.* **2015**, *27*, 2355–2360. [CrossRef]
42. Lin, Y.; Cooper, C.; Wang, M.; Adams, J.J.; Genzer, J.; Dickey, M.D. Handwritten, soft circuit boards and antennas using liquid metal nanoparticles. *Small* **2015**, *11*, 6397–6403. [CrossRef] [PubMed]
43. Tang, S.Y.; Mitchell, D.R.; Zhao, Q.; Yuan, D.; Yun, G.; Zhang, Y.; Qiao, R.; Lin, Y.; Dickey, M.D.; Li, W. Phase separation in liquid metal nanoparticles. *Matter* **2019**, *1*, 192–204. [CrossRef]

44. Ferreira, P.; Carvalho, A.; Correia, T.R.; Antunes, B.P.; Correia, I.J.; Alves, P. Functionalization of polydimethylsiloxane membranes to be used in the production of voice prostheses. *Sci. Technol. Adv. Mater.* **2013**, *14*, 055006. [CrossRef] [PubMed]
45. Pan, C.; Kumar, K.; Li, J.; Markvicka, E.J.; Herman, P.R.; Majidi, C. Visually Imperceptible Liquid-Metal Circuits for Transparent, Stretchable Electronics with Direct Laser Writing. *Adv. Mater.* **2018**, *30*, 1706937. [CrossRef] [PubMed]
46. Burakowski, T.; Wierzchon, T. *Surface Engineering of Metals: Principles, Equipment, Technologies*; CRC Press: Florida, FL, USA, 1998; pp. 333–334.
47. Orita, M.; Ohta, H.; Hirano, M.; Hosono, H. Deep-ultraviolet transparent conductive β-Ga_2O_3 thin films. *Appl. Phys. Lett.* **2000**, *77*, 4166–4168. [CrossRef]

Publisher's Note: MDPI stays neutral with regard to jurisdictional claims in published maps and institutional affiliations.

Article

Crack-Assisted Charge Injection into Solvent-Free Liquid Organic Semiconductors via Local Electric Field Enhancement

Kyoung-Hwan Kim [1], Myung-June Park [1,2] and Ju-Hyung Kim [1,2,*]

1   Department of Energy Systems Research, Ajou University, Suwon 16499, Korea;
    kkh0217@ajou.ac.kr (K.-H.K.); mjpark@ajou.ac.kr (M.-J.P.)
2   Department of Chemical Engineering, Ajou University, Suwon 16499, Korea
*   Correspondence: juhyungkim@ajou.ac.kr

Received: 26 June 2020; Accepted: 27 July 2020; Published: 28 July 2020

Abstract: Non-volatile liquid organic semiconducting materials have received much attention as emerging functional materials for organic electronic and optoelectronic devices due to their remarkable advantages. However, charge injection and transport processes are significantly impeded at interfaces between electrodes and liquid organic semiconductors, resulting in overall lower performance compared to conventional solid-state electronic devices. Here we successfully demonstrate efficient charge injection into solvent-free liquid organic semiconductors via cracked metal structures with a large number of edges leading to local electric field enhancement. For this work, thin metal films on deformable polymer substrates were mechanically stretched to generate cracks on the metal surfaces in a controlled manner, and charge injection properties into a typical non-volatile liquid organic semiconducting material, (9-2-ethylhexyl)carbazole (EHCz), were investigated in low bias region (i.e., ohmic current region). It was found that the cracked structures significantly increased the current density at a fixed external bias voltage via the local electric field enhancement, which was strongly supported by field intensity calculation using COMSOL Multiphysics software. We anticipate that these results will significantly contribute to the development and further refinement of various organic electronic and optoelectronic devices based on non-volatile liquid organic semiconducting materials.

Keywords: organic electronics; liquid semiconductors; charge injection; surface engineering; crack engineering

1. Introduction

Non-volatile liquid organic semiconducting materials have attracted a lot of interest as emerging functional materials for organic electronic and optoelectronic devices in recent years, because these fluidic materials present outstanding advantages such as tunable optoelectronic responses, degradation-free characteristics, solvent-free processability, and ultimate mechanical flexibility and uniformity [1–8]. Various electronic and optoelectronic applications, such as photorefractive devices, organic light-emitting diodes, dye-sensitized solar cells, memory devices, and optically pumped lasers, have been already demonstrated using non-volatile liquid organic semiconductors as active materials [5–16]. In particular, it has been reported that fresh liquid organic semiconductors can be continuously injected into the devices through microfluidic channels for preventing performance degradation [15,16]. However, due to the relatively low efficiency in comparison with conventional solid-state devices, considerable research efforts have been devoted to the development of novel molecular structures and device architectures to improve device performance, which is still a challenge [1,11,13–17].

Since charge carriers (i.e., electrons and holes) for operating organic electronic and optoelectronic devices should be essentially injected from electrodes, efficient charge injection into the liquid organic layers is necessarily required to realize high-performance devices based on the non-volatile liquid organic semiconductors [3,17,18]. However, the fluidity of the liquid organic materials causes difficulties in inducing preferable molecular orientations on the electrodes for facilitating charge injection, and intermolecular distances in liquid phases are intrinsically longer than those in solid phases (i.e., less dense molecular packing in liquid phases). Charge injection and transport processes are thus significantly impeded at the interfaces between the electrodes and the liquid organic materials, resulting in overall lower performance [19–25]. In this context, various methods such as inserting buffer layers and adding ionic dopants have been employed to reduce charge injection barriers and to increase charge carrier concentrations [12,26–32]. In addition, the charge carrier injection into the organic liquid materials is expected to be significantly improved if the local electric fields inducing the migration of charge carriers near the interfaces are enhanced [18,26,27,33,34]. The electric fields are spontaneously concentrated at the edges of the field plates (i.e., edge effect) [35–38], where the field intensities are locally increased, and thus the field plate structures (i.e., shapes of electrodes) play a decisive role in the spatial distributions of electric fields. This phenomenon is also applicable to interface engineering for improving the charge injection characteristics of organic electronic devices.

Here we successfully demonstrate efficient charge injection into solvent-free liquid organic semiconductors via cracked metal structures with a large number of edges leading to local electric field enhancement. For this work, silver (Ag) thin films, deposited on deformable fluorinated ethylene propylene (FEP) substrates, were used to generate cracks on the field plates in a controlled manner. The Ag films on the polymer substrates were mechanically gripped and stretched up to fixed ratios, resulting in the formation of the cracks with reproducible patterns [39,40]. Although this simple cracking method can be easily performed to fabricate a large number of edges on metal surfaces without any lithographic processes, heavily cracked metal electrodes normally give a rise to high electrical resistance resulting in performance degradation [41–43]. To avoid increases in electrical resistance originating from the structural deformations, the cracked Ag films were transferred and welded onto other intact Ag films to complete the electrode structures, and then the charge injection properties were investigated using a typical non-volatile liquid organic semiconducting material, (9-2-ethylhexyl)carbazole (EHCz) [1,5–8,10,12–14]. The fluidity of EHCz with a glass transition temperature below 0 °C facilitates the penetration of the molecules into the crack structures of the electrodes, clearly showing the effects of the engineered interfacial structures in charge injection. It was found that the cracked structures significantly increased the current density at a fixed external bias voltage via the local electric field enhancement, which was strongly supported by field intensity calculation using COMSOL Multiphysics software. These results suggest great potential for the development and further refinement of various organic electronic and optoelectronic devices based on non-volatile liquid organic semiconducting materials.

2. Materials and Methods

2.1. Materials

The deformable FEP films with a thickness of 0.125 mm (Teflon®FEP) and EHCz were purchased from Alphaflon (Seoul, Korea) and Sigma Aldrich (Seoul, Korea), respectively, and used as received.

2.2. Device Fabrication

The cracked metal electrodes used in this work were prepared as schematically illustrated in Figure 1. The Ag thin films with a thickness of 30 nm were preferentially deposited on the deformable FEP substrates by thermal evaporation in vacuum ($<1.0 \times 10^{-6}$ Torr). The samples were individually gripped in a rectangular frame for applying tensile forces, and then uniaxially and biaxially stretched to 120% and 140%, respectively (designated "UA120", "UA140", "BA120", and "BA140"; see also

Table 1). It should be noted that the maximum stretching ratio which was reliable without tearing or slipping of the sample in our experimental setup was 140%. To prevent an increase in electrical resistance, each cracked Ag film was transferred and welded onto another intact Ag film by means of cold-welding [44–47]. For the cold-welding process, the stretched Ag/FEP sample was brought into contact with the intact Ag film (with a 30-nm thickness) deposited on a glass substrate, and subsequently pressed with a pressure of 0.2 MPa for 90 s at room temperature. The FEP substrate was then easily peeled off from the sample without residue owing to a lower surface energy of FEP (see Figure 1c). All the transferred samples exhibited no significant change in high electrical conductivity, compared to a reference electrode (i.e., the intact Ag film with a 60-nm thickness deposited on the glass substrate). It was also confirmed that the cracked Ag films were neatly transferred onto the intact Ag films, using scanning electron microscopy (SEM).

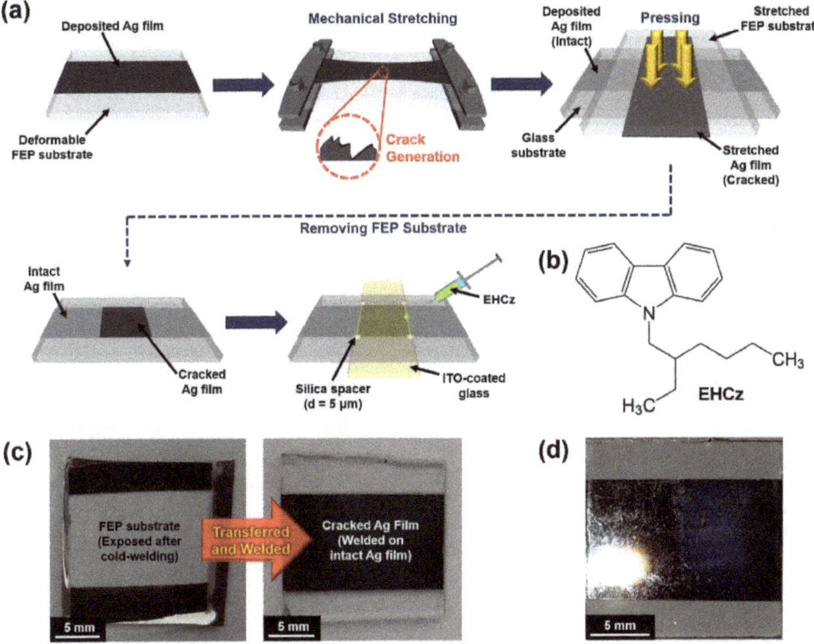

Figure 1. (**a**) Schematic illustration of the preparation of cracked Ag electrode in a controlled manner, and the fabrication of device. (**b**) Chemical structure of (9-2-ethylhexyl)carbazole (EHCz). (**c**) Photographic images of the cracked Ag film after completing the cold-welding process. (**d**) Photographic image of the cracked Ag film inducing diffuse reflection.

Table 1. Stretching ratios of Ag films.

Sample Name	Stretching Ratio (X-Axis)	Stretching Ratio (Y-Axis)
Reference	0%	0%
UA120	20%	0%
UA140	40%	0%
BA120	20%	20%
BA140	40%	40%

X- and Y-axes are in-plane, which are perpendicular to each other.

Each prepared Ag electrode was covered with another glass substrate coated with indium-tin oxide (ITO) (20 Ω sq^{-1}), and silica microsphere spacers (of 5-μm diameter) were used for a fixed gap distance between Ag and ITO. The gap between the two electrodes was then filled with EHCz by capillary action to complete the device structure (see Figure 2). The active area of each device was 1×1 cm^2. It is worth noting that EHCz is highly viscous, which substantially hinders the infiltration into smaller gaps (in the submicron range). Even if EHCz well stays in the gap without leakage due to its high viscosity, the two substrates (i.e., lower and cover glasses) should be securely fixed to prevent slipping by the liquid.

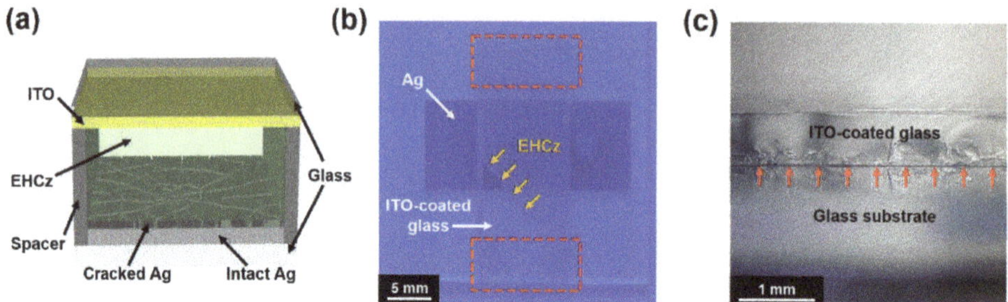

Figure 2. (**a**) Configuration of device to investigate the charge injection properties via local electric field enhancement. (**b**) Photographic image of the device illustrated in (**a**), under 365 nm UV light. The optically excited EHCz material under UV irradiation clearly revealed its liquid boundary during the capillary action (indicated by yellow arrows). Red dashed boxes indicate the locations of silica microsphere spacers. (**c**) Optical microscopy image of a cross section of the device in (**b**). The gap between the two electrodes is indicated by red arrows.

2.3. Measurements

The current density-voltage (J-V) characteristics of the devices were measured in response to a voltage sweep from 0.0 to +1.0 V, using a Keithley 2636 source meter. The capacitance of EHCz was also measured with Agilent 4284A Precision LCR Meter (Agilent, Santa Clara, CA, USA), to evaluate the dielectric constant.

2.4. Field Intensity Calculation

The field intensity within the device was calculated using COMSOL Multiphysics software (Burlington, MA, USA). Model structure of the device was prepared, based on the experimental observation as will be discussed in the Section 3. Static analysis was also performed to elucidate the spatial distribution of local electric fields within the device at a fixed bias voltage of +1.0 V.

3. Results

The prepared Ag electrodes via mechanical stretching exhibited a matt dark gray color, because the cracked metal structures cause diffuse reflections rather than specular reflections (see Figure 1d). To confirm the overall shapes of the cracks formed on the Ag electrodes, the samples after completing the cold-welding processes were investigated using SEM as shown in Figures 3 and 4. For the UA120 and UA140 samples (i.e., uniaxially stretched to 120% and 140%, respectively), the Ag films tended to crack in the form of a one-dimensional line. With an increase in the uniaxial stretching ratio, the overall crack size increased, and the edges of the cracks were more clearly observed. Particularly, in the UA140 sample, minor cracks were further observed between the major cracks, leading to a higher crack density compared to UA120 (see Figure 3d). For the BA120 sample, which was biaxially stretched to 120%, the orientations of the cracks were significantly diversified in comparison with the uniaxially stretched samples. The overall shape and density of the cracks were comparable to the minor cracks of UA140; however, large cracks similar to the major cracks of UA140 with the well-defined edges were not observed in BA120. As the biaxial stretching ratio increased up to 140% (i.e., BA140), the Ag film was eventually divided into island forms by major cracks, and an average area of the islands was found to be less than 3 μm². In comparison with the other samples, minor cracks with a higher density were clearly observed within the Ag islands as shown in Figure 3h.

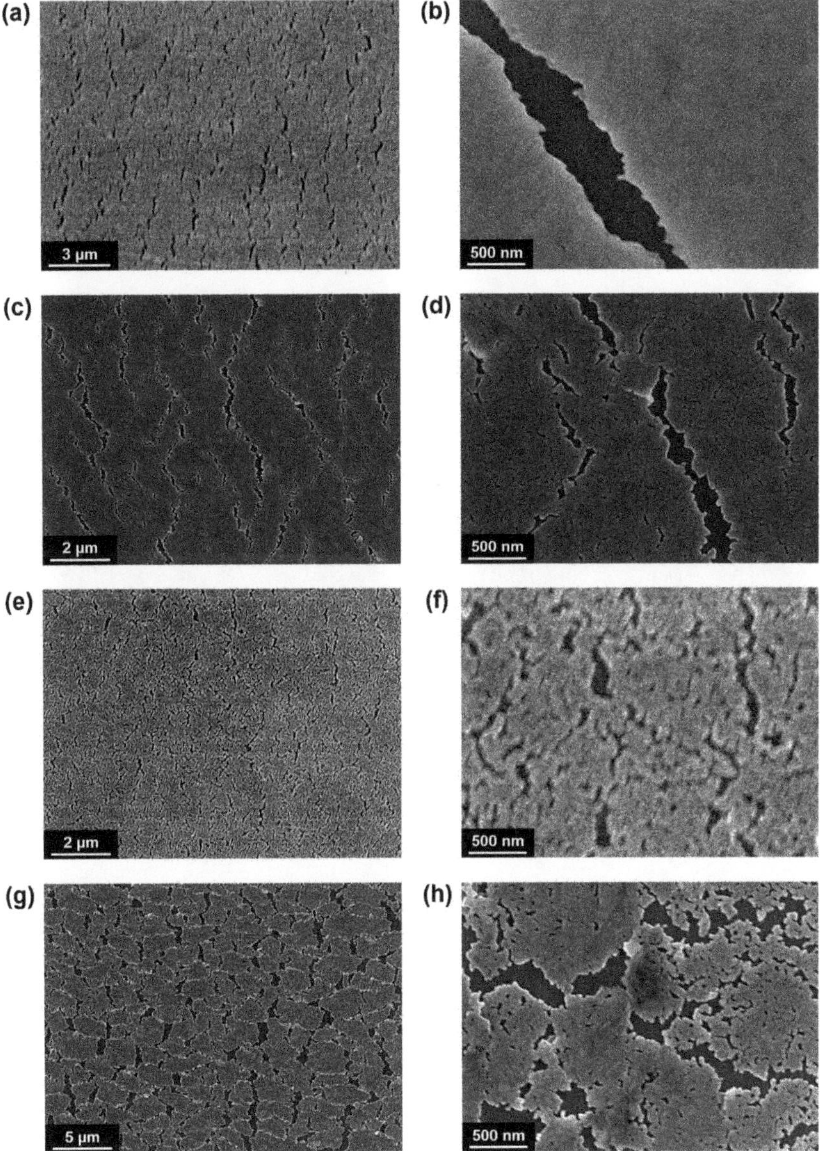

Figure 3. SEM images of the cracked Ag electrodes after completing the cold-welding processes. (**a,b**) SEM images of the UA120 sample. No further deformation of Ag was observed between crack lines as in (**b**). (**c,d**) SEM images of the UA140 samples. Minor cracks were further observed between major crack lines as in (**d**). (**e,f**) SEM images of the BA120 sample. Orientations of cracks were significantly diversified in comparison with the uniaxially stretched samples as in (**f**). (**g,h**) SEM images of the BA140 sample. The Ag film was divided into island forms as in (**g**), and minor cracks with a high density were clearly observed within the Ag islands as in (**h**).

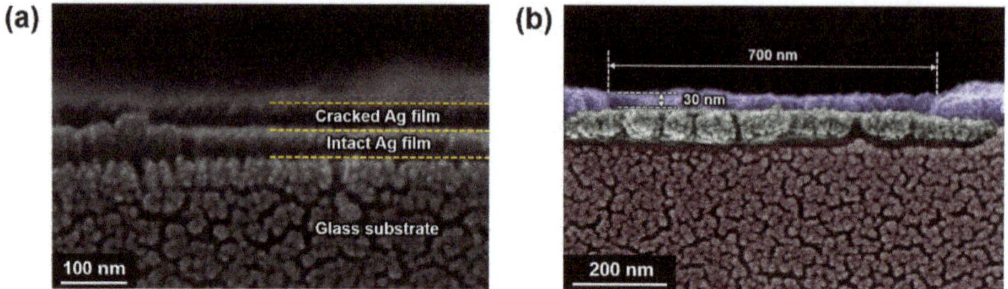

Figure 4. (**a**) SEM image of a selected cross section of the BA140 sample after completing the cold-welding process. (**b**) Colored SEM image of a selected cross section of the same sample as in (**a**). Red, green, and blue regions represent the glass substrate, the intact Ag film, and the cracked Ag film, respectively. The cross section of the crack (with 700 nm width and 30 nm depth) is clearly revealed in the blue region.

Figure 4 also shows cross-sectional SEM images of the BA140 sample. For these SEM measurements, platinum (Pt) was coated onto the sample with a thickness of ~5 nm to clearly observe the non-conductive glass substrate. It was confirmed that the cracked Ag film was neatly transferred and welded on the intact Ag film through the cold-welding process. It is worth noting here that metal oxide layers could not be easily transferred and welded onto other substrates in our experimental setup, and thus binding materials would be required to enhance adhesion.

The J-V characteristics for the devices were measured in response to a voltage sweep from 0.0 to +1.0 V, as shown in Figure 5. It should be noted that the charge carriers are normally accumulated in the organic layers in high bias voltage region, due to the relatively low charge carrier mobility. Such charge carrier accumulations in the organic layers give a rise to changes not only in the electric field distributions, but also in the J-V characteristics (i.e., from ohmic currents to space-charge-limited currents) [17,26,28,48,49]. Thus, changes in the charge injection properties according to the local electric field enhancement can be more clearly examined in low bias region (i.e., ohmic current region). As indicated in Figure 5, the current density was gradually increased as the density of the cracks on the Ag electrode increased. In particular, it was found that the slope ratio of the device with the BA140 electrode to the reference devices was ~170 in the J-V characteristics. At a fixed bias voltage of +1.0 V, the current densities were measured to be 6.68×10^{-8}, 6.60×10^{-7}, 1.08×10^{-6}, 2.20×10^{-6}, and 1.16×10^{-5} A cm^{-2} for reference, UA120, BA120, UA140, and BA140, respectively. It is worth noting that EHCz is an intrinsically p-type material, of which conduction is entirely governed by holes [1], and the devices presented in this work can be described as hole-only devices.

All the devices showed linear J-V characteristics within the bias voltage range (i.e., 0.0 to +1.0 V), indicating the ohmic behaviors of the devices in low bias region. If there is no change in charge carrier mobility, the ohmic currents of organic electronic devices at constant temperature are normally enhanced as the initial concentrations of charge carriers within the semiconducting layers increase using ionic dopants [31,32]. However, in this work, the increases in the ohmic currents were solely induced by the injected charges from the electrodes without using any extra dopants. These results strongly suggest that efficient charge injection via local electric field enhancement can exert similar effects to the introduction of ionic dopants on J-V characteristics in terms of charge carrier concentrations. It is notable that the maximum ohmic current density of 1.16×10^{-5} A cm^{-2} in this work is relatively lower than those of other solid-state devices based on carbazole derivatives or similar organic semiconducting materials. In ITO/undoped organic semiconductor/metal structures, previous works reported ohmic current densities of ~10^{-4} A cm^{-2} for 4,4′,4′′-tris(N-3-methylphenyl-N- phenyl-amino)-triphenylamine [32], poly(2,6-diphenyl-4-((9-ethyl)-9*H*-carbazole)-pyridinyl-*alt*-2,7-(9,9-didodecyl)-9*H*-fluorenyl) [50], and N,N′-bis(3-methylphenyl)-N,N′-diphenylbenzidine [51], and even higher ohmic current densities were also found with a few of carbazole derivatives [52]. In addition, a few photoluminescent devices were successfully demonstrated using EHCz as a host material in the previous studies [5,6,8]; however, at the current stage, we could not observe electroluminescence from dye-doped EHCz materials due to the imbalance of electrons and holes.

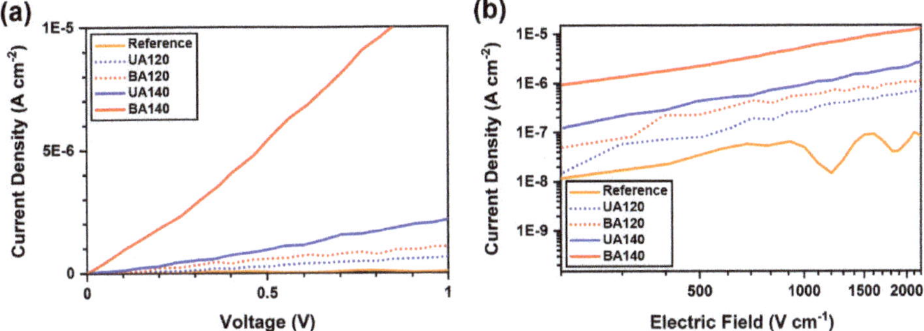

Figure 5. (**a**) Current density-voltage (J-V) characteristics of the devices using the reference, UA120, UA140, BA120, and BA140 electrodes, in a linear scale. The characteristics were measured in low bias region (i.e., ohmic current region). All the characteristics showed a linear relationship, and that the slope ratio of the device with the BA140 electrode to the reference devices was ~170. (**b**) Electric field dependence of the current density. All the slopes were measured to be ~1, indicating ohmic current in each device.

The field intensity calculation was also performed to investigate the local electric field enhancement induced by a highly cracked metal structure as shown in Figure 6. The cracked Ag structure with a depth of 30 nm was considered for the calculation, on the basis of the observed SEM image (see Figure 6b). In the simulation model, the gap distance between the lower Ag and upper ITO electrodes was fixed at 5 μm, and the gap was filled with a dielectric material corresponding to EHCz. For the dielectric material, dielectric constant of 3.02, electrical conductivity of 2.1×10^{-9} S cm^{-1}, and density of 1.004 g mL^{-1} were used as material parameters to simulate EHCz. It should be noted that the dielectric constant of EHCz was experimentally measured for this work. For the measurement, EHCz was injected into the gap between two Ag electrodes with a gap distance of 5 μm, of which the capacitance was monitored at a frequency of 1 kHz. The measured capacitance was converted into the dielectric constant in consideration of active area and thickness.

Static analysis was performed to clarify the spatial distribution of local electric fields within the device at a fixed bias voltage of +1.0 V. According to the distance from the Ag bottom, the local electric field intensities were calculated for the selected cross sections, as shown in Figure 6a. The local electric field intensities were significantly enhanced within the cracks of the Ag electrode, of which apexes were found at the upper edges. In particular, the intensity of the local electric field increased up to ~4000 V cm^{-1} at the upper edges. To clearly show the local electric field enhancement, the field intensities for the selected cross section were further visualized as shown in Figure 6c. These calculation results were in good agreement with the experimental results, where charge injection properties were dramatically improved by the high-density cracks contributing to local electric field enhancement.

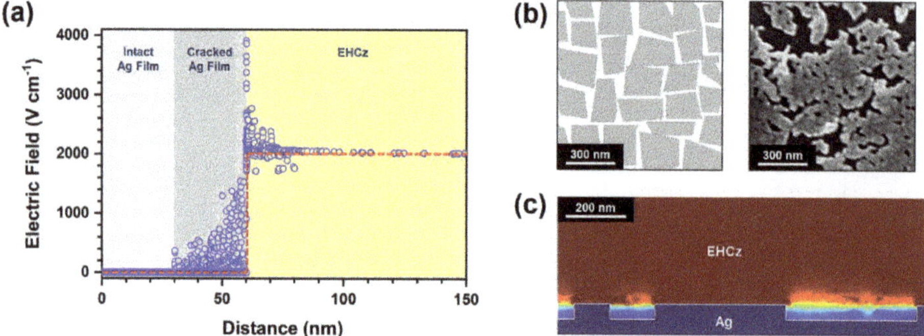

Figure 6. (**a**) Local electric field intensities according to the distance from the Ag bottom, based on the field intensity calculation at a fixed bias voltage of +1.0 V. Blue circles indicate the local electric field intensities in the device with a highly cracked Ag structure (30-nm depth), and a red dashed line indicates the electric field in the reference device. (**b**) Model structure configured for the field intensity calculation in (**a**), and SEM image of BA140 used to prepare the model structure. (**c**) Local electric field intensities visualized for a selected cross section. The boundary between Ag and EHCz is indicated by a white line. The intensity level, presented by colors, increases from blue to red (i.e., blue-green-yellow-red).

4. Summary

We demonstrate efficient charge injection into solvent-free liquid organic semiconductors via cracked metal structures with a large number of edges leading to local electric field enhancement. For this work, the Ag thin films on the deformable FEP substrates were mechanically stretched to generate the cracks on the surfaces in a controlled manner, and subsequently transferred and welded onto other intact Ag films for avoiding increases in electrical resistance. Using the prepared Ag electrodes with varied crack densities, the charge injection properties into EHCz were investigated in low bias region (i.e., ohmic current region). It was found that the device with the highly cracked electrode dramatically increased the current density within the ohmic current region, indicating that efficient charge injection via local electric field enhancement can exert similar effects to the introduction of ionic dopants on J-V characteristics in terms of charge carrier concentrations. Although the hole-only devices were demonstrated in this work, these results still offer a wide range of possibilities for various device applications. Careful consideration of work functions of electrodes and cascade energy levels would be required for optimizing the device design or selection of device components. In addition, the field intensity within the device was calculated using COMSOL Multiphysics software, based on the experimental observation, and static analysis was also performed to reveal the spatial distribution of local electric fields. The calculation results were in good agreement with our experimental results, where charge injection properties were dramatically improved by the high-density cracks contributing to local electric field enhancement. We anticipated that these results will significantly contribute to the development and further refinement of various organic electronic and optoelectronic devices based on non-volatile liquid organic semiconducting materials.

Author Contributions: K.-H.K.: methodology, software, formal analysis, investigation, data curation, visualization, and writing—original draft preparation; M.-J.P.: methodology, software, resources, and supervision; J.-H.K.: conceptualization, methodology, validation, writing—original draft preparation, review and editing, supervision, project administration, funding acquisition. All authors have read and agreed to the published version of the manuscript.

Funding: This research was funded by the Basic Science Research Program through the National Research Foundation of Korea (NRF-2018R1C1B6003122).

Conflicts of Interest: The authors declare no conflict of interest.

References

1. Ribierre, J.-C.; Aoyama, T.; Muto, T.; Imase, Y.; Wada, T. Charge transport properties in liquid carbazole. *Org. Electron.* **2008**, *9*, 396–400. [CrossRef]
2. Schmidt, H.; Hawkins, A.R. The photonic integration of non-solid media using optofluidics. *Nat. Photon.* **2011**, *5*, 598–604. [CrossRef]
3. Kamino, B.A.; Bender, T.P.; Klenkler, R.A. Hole mobility of a liquid organic semiconductor. *J. Phys. Chem. Lett.* **2012**, *3*, 1002–1006. [CrossRef]
4. Babu, S.S.; Aimi, J.; Ozawa, H.; Shirahata, N.; Saeki, A.; Seki, S.; Ajayaghosh, A.; Möhwald, H.; Nakanishi, T. Solvent-free luminescent organic liquids. *Angew. Chem. Int. Ed. Engl.* **2012**, *51*, 3391–3395. [CrossRef]
5. Choi, E.Y.; Mager, L.; Cham, T.T.; Dorkenoo, K.D.; Fort, A.; Wu, J.W.; Barsella, A.; Ribierre, J.-C. Solvent-free fluidic organic dye lasers. *Opt. Express* **2013**, *21*, 11368–11375. [CrossRef]
6. Kim, J.-H.; Inoue, M.; Zhao, L.; Komino, T.; Seo, S.; Ribierre, J.-C.; Adachi, C. Tunable and flexible solvent-free liquid organic distributed feedback lasers. *Appl. Phys. Lett.* **2015**, *106*, 053302. [CrossRef]
7. Ribierre, J.C.; Zhao, L.; Inoue, M.; Schwartz, P.O.; Kim, J.H.; Yoshida, K.; Sandanayaka, A.S.; Nakanotani, H.; Mager, L.; Mery, S.; et al. Low threshold amplified spontaneous emission and ambipolar charge transport in non-volatile liquid fluorene derivatives. *Chem. Commun.* **2016**, *52*, 3103–3106. [CrossRef] [PubMed]
8. Sandanayaka, A.S.D.; Zhao, L.; Pitrat, D.; Mulatier, J.-C.; Matsushima, T.; Andraud, C.; Kim, J.-H.; Ribierre, J.-C.; Adachi, C. Improvement of the quasi-continuous-wave lasing properties in organic semiconductor lasers using oxygen as triplet quencher. *Appl. Phys. Lett.* **2016**, *108*, 223301. [CrossRef]
9. Snaith, H.J.; Zakeeruddin, S.M.; Wang, Q.; Péchy, P.; Grätzel, M. Dye-sensitized solar cells incorporating a liquid hole-transporting material. *Nano Lett.* **2006**, *6*, 2000–2003. [CrossRef]
10. Ribierre, J.-C.; Aoyama, T.; Kobayashi, T.; Sassa, T.; Muto, T.; Wada, T. Influence of the liquid carbazole concentration on charge trapping in C_{60} sensitized photorefractive polymers. *J. Appl. Phys.* **2007**, *102*, 033106. [CrossRef]
11. Xu, D.; Adachi, C. Organic light-emitting diode with liquid emitting layer. *Appl. Phys. Lett.* **2009**, *95*, 053304. [CrossRef]
12. Ribierre, J.-C.; Aoyama, T.; Muto, T.; André, P. Hybrid organic–inorganic liquid bistable memory devices. *Org. Electron.* **2011**, *12*, 1800–1805. [CrossRef]
13. Hirata, S.; Kubota, K.; Jung, H.H.; Hirata, O.; Goushi, K.; Yahiro, M.; Adachi, C. Improvement of electroluminescence performance of organic light-emitting diodes with a liquid-emitting layer by introduction of electrolyte and a hole-blocking layer. *Adv. Mater.* **2011**, *23*, 889–893. [CrossRef]
14. Kubota, K.; Hirata, S.; Shibano, Y.; Hirata, O.; Yahiro, M.; Adachi, C. Liquid carbazole substituted with a poly(ethylene oxide) group and its application for liquid organic light-emitting diodes. *Chem. Lett.* **2012**, *41*, 934–936. [CrossRef]
15. Shim, C.-H.; Hirata, S.; Oshima, J.; Edura, T.; Hattori, R.; Adachi, C. Uniform and refreshable liquid electroluminescent device with a back side reservoir. *Appl. Phys. Lett.* **2012**, *101*, 113302. [CrossRef]
16. Kasahara, T.; Matsunami, S.; Edura, T.; Oshima, J.; Adachi, C.; Shoji, S.; Mizuno, J. Fabrication and performance evaluation of microfluidic organic light emitting diode. *Sensor. Actuat. A Phys.* **2013**, *195*, 219–223. [CrossRef]
17. Plint, T.G.; Kamino, B.A.; Bender, T.P. Charge carrier mobility of siliconized liquid triarylamine organic semiconductors by time-of-flight spectroscopy. *J. Phys. Chem. C* **2015**, *4*, 1676–1682. [CrossRef]
18. Coropceanu, V.; Cornil, J.; da Silva Filho, D.A.; Olivier, Y.; Silbey, R.; Brédas, J.-L. Charge transport in organic semiconductors. *Chem. Rev.* **2007**, *107*, 926–952. [CrossRef]
19. Forrest, S.R. Ultrathin organic films grown by organic molecular beam deposition and related techniques. *Chem. Rev.* **1997**, *97*, 1793–1896. [CrossRef]
20. Ishii, H.; Sugiyama, K.; Ito, E.; Seki, K. Energy level alignment and interfacial electronic structures at organic/metal and organic/organic interfaces. *Adv. Mater.* **1999**, *11*, 605–625. [CrossRef]
21. Hooks, D.E.; Fritz, T.; Ward, M.D. Epitaxy and molecular organization on solid substrates. *Adv. Mater.* **2001**, *13*, 227–241. [CrossRef]
22. Zhu, X.-Y. Electronic structure and electron dynamics at molecule-metal interfaces: Implications for molecule-based electronics. *Surf. Sci. Rep.* **2004**, *56*, 1–83. [CrossRef]

23. Yamane, H.; Kanai, K.; Ouchi, Y.; Ueno, N.; Seki, K. Impact of interface geometric structure on organic–metal interface energetics and subsequent films electronic structure. *J. Electron Spectrosc. Relat. Phenom.* **2009**, *174*, 28–34. [CrossRef]

24. Ma, H.; Yip, H.-L.; Huang, F.; Jen, A.K.-Y. Interface engineering for organic electronics. *Adv. Funct. Mater.* **2010**, *20*, 1371–1388. [CrossRef]

25. Kim, J.-H. Interfacial phenomena between conjugated organic molecules and noble metals. *Korean J. Chem. Eng.* **2017**, *34*, 1281–1293. [CrossRef]

26. Jain, S.; Geens, W.; Mehra, A.; Kumar, V.; Aernouts, T.; Poortmans, J.; Mertens, R.; Willander, M. Injection-and space charge limited-currents in doped conducting organic materials. *J. Appl. Phys.* **2001**, *89*, 3804–3810. [CrossRef]

27. Hosseini, A.; Wong, M.H.; Shen, Y.; Malliaras, G.G. Charge injection in doped organic semiconductors. *J. Appl. Phys.* **2005**, *97*, 023705. [CrossRef]

28. Matsushima, T.; Kinoshita, Y.; Murata, H. Formation of ohmic hole injection by inserting an ultrathin layer of molybdenum trioxide between indium tin oxide and organic hole-transporting layers. *Appl. Phys. Lett.* **2007**, *91*, 253504. [CrossRef]

29. Kim, J.; Khang, D.-Y.; Kim, J.-H.; Lee, H.H. The surface engineering of top electrode in inverted polymer bulk-heterojunction solar cells. *Appl. Phys. Lett.* **2008**, *92*, 133307. [CrossRef]

30. Kim, J.-H.; Huh, S.-Y.; Kim, T.-i.; Lee, H.H. Thin pentacene interlayer for polymer bulk-heterojunction solar cell. *Appl. Phys. Lett.* **2008**, *93*, 143305. [CrossRef]

31. Lee, J.-H.; Leem, D.-S.; Kim, H.-J.; Kim, J.-J. Effectiveness of p-dopants in an organic hole transporting material. *Appl. Phys. Lett.* **2009**, *94*, 123306. [CrossRef]

32. Shan, M.; Jiang, H.; Guan, Y.; Sun, D.; Wang, Y.; Hua, J.; Wang, J. Enhanced hole injection in organic light-emitting diodes utilizing a copper iodide-doped hole injection layer. *RSC Adv.* **2017**, *7*, 13584–13589. [CrossRef]

33. Arkhipov, V.; Emelianova, E.; Tak, Y.; Bässler, H. Charge injection into light-emitting diodes: Theory and experiment. *J. Appl. Phys.* **1998**, *84*, 848–856. [CrossRef]

34. Mahapatro, A.K.; Ghosh, S. Schottky energy barrier and charge injection in metal/copper–phthalocyanine/metal structures. *Appl. Phys. Lett.* **2002**, *80*, 4840–4842. [CrossRef]

35. Fowler, R.H.; Nordheim, L. Electron emission in intense electric fields. *Proc. R. Soc. A* **1928**, *119*, 173–181.

36. De Heer, W.A.; Chatelain, A.; Ugarte, D. A carbon nanotube field-emission electron source. *Science* **1995**, *270*, 1179–1180. [CrossRef]

37. Fujii, S.; Honda, S.-i.; Machida, H.; Kawai, H.; Ishida, K.; Katayama, M.; Furuta, H.; Hirao, T.; Oura, K. Efficient field emission from an individual aligned carbon nanotube bundle enhanced by edge effect. *Appl. Phys. Lett.* **2007**, *90*, 153108. [CrossRef]

38. Li, Y.; Sun, Y.; Yeow, J.T. Nanotube field electron emission: Principles, development, and applications. *Nanotechnology* **2015**, *26*, 242001. [CrossRef]

39. Ye, T.; Suo, Z.; Evans, A. Thin film cracking and the roles of substrate and interface. *Int. J. Solids Struct.* **1992**, *29*, 2639–2648. [CrossRef]

40. Li, T.; Suo, Z. Deformability of thin metal films on elastomer substrates. *Int. J. Solids Struct.* **2006**, *43*, 2351–2363. [CrossRef]

41. Kang, D.; Pikhitsa, P.V.; Choi, Y.W.; Lee, C.; Shin, S.S.; Piao, L.; Park, B.; Suh, K.-Y.; Kim, T.-i.; Choi, M. Ultrasensitive mechanical crack-based sensor inspired by the spider sensory system. *Nature* **2014**, *516*, 222–226. [CrossRef] [PubMed]

42. Park, B.; Kim, J.; Kang, D.; Jeong, C.; Kim, K.S.; Kim, J.U.; Yoo, P.J.; Kim, T.-i. Dramatically enhanced mechanosensitivity and signal-to-noise ratio of nanoscale crack-based sensors: Effect of crack depth. *Adv. Mat.* **2016**, *28*, 8130–8137. [CrossRef] [PubMed]

43. Kim, M.; Choi, H.; Kim, T.; Hong, I.; Roh, Y.; Park, J.; Seo, S.; Han, S.; Koh, J.-s.; Kang, D. FEP encapsulated crack-based sensor for measurement in moisture-laden environment. *Materials* **2019**, *12*, 1516. [CrossRef] [PubMed]

44. Bay, N. Cold pressure welding—The mechanisms governing bonding. *J. Eng. Ind.* **1979**, *101*, 121–127. [CrossRef]

45. Kim, C.; Burrows, P.E.; Forrest, S.R. Micropatterning of organic electronic devices by cold-welding. *Science* **2000**, *288*, 831–833. [CrossRef]

46. Kim, C.; Shtein, M.; Forrest, S.R. Nanolithography based on patterned metal transfer and its application to organic electronic devices. *Appl. Phys. Lett.* **2002**, *80*, 4051–4053. [CrossRef]

47. Asare, J.; Adeniji, S.; Oyewole, O.; Agyei-Tuffour, B.; Du, J.; Arthur, E.; Fashina, A.; Zebaze Kana, M.; Soboyejo, W. Cold welding of organic light emitting diode: Interfacial and contact models. *AIP Adv.* **2016**, *6*, 065125. [CrossRef]

48. Varo, P.L.; Tejada, J.J.; Villanueva, J.L.; Carceller, J.; Deen, M. Modeling the transition from ohmic to space charge limited current in organic semiconductors. *Org. Electron.* **2012**, *13*, 1700–1709. [CrossRef]

49. Wetzelaer, G.; Blom, P.W. Ohmic current in organic metal-insulator-metal diodes revisited. *Phys. Rev. B* **2014**, *89*, 241201. [CrossRef]

50. Liu, B.G.; Liaw, D.-J.; Lee, W.-Y.; Ling, Q.-D.; Zhu, C.-X.; Chan, D.S.-H.; Kang, E.-T.; Neoh, K.-G. Tristable electrical conductivity switching in a polyfluorene-diphenylpyridine copolymer with pendant carbazole groups. *Phil. Trans. R. Soc. A* **2009**, *367*, 4203–4214. [CrossRef]

51. Havare, A.K. Effect of the interface improved by self-assembled aromatic organic semiconductor molecules on performance of OLED. *ECS J. Solid State Sci. Technol.* **2020**, *9*, 041007. [CrossRef]

52. Noine, K.; Pu, Y.-J.; Nakayama, K.-i.; Kido, J. Bifluorene compounds containing carbazole and/or diphenylamine groups and their bipolar charge transport properties in organic light emitting devices. *Org. Electron.* **2010**, *11*, 717–723. [CrossRef]

Article

The Methods and Experiments of Shape Measurement for Off-Axis Conic Aspheric Surface

Shijie Li [1,2,*], Jin Zhang [1,2], Weiguo Liu [1,2], Haifeng Liang [1,2], Yi Xie [3] and Xiaoqin Li [4]

[1] Shaanxi Province Key Laboratory of Thin Film Technology and Optical Test, Xi'an Technological University, Xi'an 710021, China; zhangjin@xatu.edu.cn (J.Z.); wgliu@163.com (W.L.); hfliang2004@163.com (H.L.)
[2] School of Optoelectronic Engineering, Xi'an Technological University, Xi'an 710021, China
[3] Display Technology Department, Luoyang Institute of Electro-optical Equipment, AVIC, Luoyang 471009, China; xyi613@163.com
[4] Library of Xi'an Technological University, Xi'an Technological University, Xi'an 710021, China; rileyhome923@126.com
* Correspondence: lishijie@xatu.edu.cn

Received: 26 March 2020; Accepted: 27 April 2020; Published: 1 May 2020

Abstract: The off-axis conic aspheric surface is widely used as a component in modern optical systems. It is critical for this kind of surface to obtain the real accuracy of the shape during optical processing. As is widely known, the null test is an effective method to measure the shape accuracy with high precision. Therefore, three shape measurement methods of null test including auto-collimation, single computer-generated hologram (CGH), and hybrid compensation are presented in detail in this research. Although the various methods have their own advantages and disadvantages, all methods need a special auxiliary component to accomplish the measurement. In the paper, an off-axis paraboloid (OAP) was chosen to be measured using the three methods along with auxiliary components of their own and it was shown that the experimental results involved in peak-to-valley (PV), root-mean-square (RMS), and shape distribution from three methods were consistent. As a result, the correctness and effectiveness of these three measurement methods were confirmed, which are very useful in engineering.

Keywords: off-axis conic surface; shape accuracy; auto-collimation; single CGH; hybrid compensation

---

1. Introduction

Off-axis aspheric optical elements are often used in modern optical systems such as off-axis TMA (Three Mirror Anastigmatic), telescope (e.g., GMT (Giant Magellan Telescope), TMT (Thirty Meter Teloscope), E-ELT (European Extremely Large Telescope)), and so on [1,2]. In order to achieve high quality, the off-axis aspheric optical element has become the focus of research in optical manufacturing. To evaluate the quality of off-axis aspheric surface, the related optical measurement technique is necessary to gain the shape accuracy, which is the key parameter of the optical element. Usually, interferometry is commonly used to test the shape accuracy of the flat and the spherical optical elements. However, until now, there has been no unified method for aspheric surfaces, especially off-axis aspheric surfaces, to test their profile with a nano-precision.

To measure the shape error of the off-axis aspheric surface at a high accuracy, researchers have offered several useful methods [3–13]. Wang Xiao-kun tested an off-axis ellipsoid mirror with sub-aperture stitching interferometry and applied least-squares fitting to process the test data, resulting in a 1.275λ of PV and 0.113λ of RMS [4]. Similarly, Yongfu Wen tested an off-axis hyperboloid mirror with off-axis annular sub-aperture stitching interferometry and obtained the results by a complex calculation [5]. Obviously, this sub-aperture stitching technology requires more measuring time, and the more complex data processing method is not a null test technique [4–6]. Jan Burke used a flat mirror as the aiding element to detect a 90° off-axis paraboloid mirror, which obtained the results of PV = 343 nm and RMS = 50 nm. Although this method belongs to the auto-collimation measurement method, the adjustment processing is difficult [7]. Similarly, Ki-Beom Ahn used a spherical convex reference mirror as the aiding element to detect the ellipsoid mirror, which was the secondary mirror of the Giant Magellan Telescope (GMT) [8]. Due to the diameter of this ellipsoid (up to 1.06 m), the aperture size of the spherical convex reference mirror would reach 0.99 m, leading to serious difficulties in the fabrication of this aiding element. Additionally, the adjustment processing is tough. As a null test technique, the computer generated hologram (CGH) method is a suitable compensator for off-axis aspheric surface measurement [9–11]. M. M. Talha measured a freeform surface with the CGH method with a result of PV = 0.0479λ [12]. Similarly, Li Fa-zhi used CGH as the aiding element to measure an off-axis high order aspheric surface [13].

When the aperture of the off-axis aspheric surface is up to the meter level and the asphericity rises to the millimeter level, the CGH method will not be suitable because the difficulty and cost of CGH fabrication increase dramatically. In this instance, the application of the two aiding elements, fold mirror together with GGH, could be a good choice. J. H. Burge used this method to measure the segment of primary mirror of the GMT, which was an off-axis aspheric surface with an aperture size of 8.4 m [14]. In the same way, Chang Jin Oh measured an off-axis paraboloid with a 4.2 m aperture diameter and 9 mm asphericity [15].

In order to obtain accurate measurement results of the off-axis conic aspheric surface, three null test methods are introduced in Section 2. After that, an off-axis paraboloid (OAP) was chosen to be tested in Section 3. The aiding elements from the three methods were designed and fabricated respectively to measure the OAP. Finally, the correctness of three methods were mutually cross-checked by their experimental results.

2. The Shape Measurement Methods

Interferometry is frequently reckoned as an effective tool to test the shape accuracy of the optical surface; however, it is limited to directly measuring flat or spherical surfaces. Due to the inherent aberration of the off-axis aspheric surfaces, an auxiliary optical element is required for the null test. Moreover, the aiding element changes with the variation of the measurement method.

Due to having perfect image points, a conic aspheric surface such as paraboloid, ellipsoid, and hyperboloid can be measured using the classical null test method of auto-collimation. Within this method, a flat mirror (for paraboloid) or a spherical mirror (for ellipsoid or hyperboloid) is required as the aiding optical element. As shown in Figure 1, two optical layouts for measuring an off-axis paraboloid with flat mirror and one layout for measuring an off-axis ellipsoid with a convex sphere mirror are illustrated.

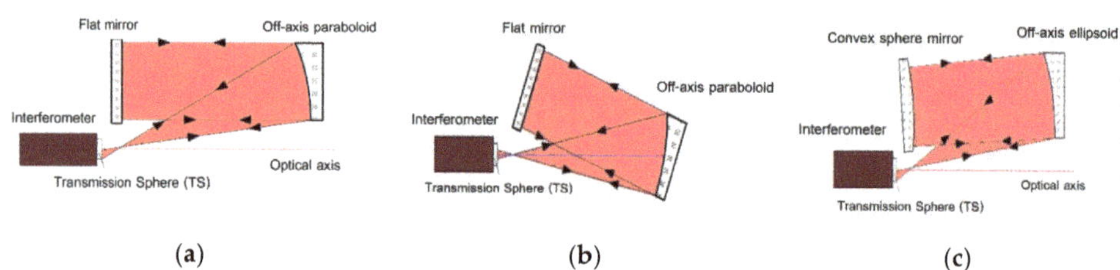

Figure 1. Schematic diagram of auto-collimation: (**a**) off-axis paraboloid located at off-axis; (**b**) off-axis paraboloid located at on-axis; (**c**) off-axis ellipsoid located at off-axis.

2.1. Auto-Collimation Method

The auto-collimation method is inexpensive and does not need a complex design or a long time to prepare. To make this method simple and useful, the aperture and the F-number of the aiding element should be considered. As shown in Equation (1), there are three errors occurring in the test results of this method.

$$Error_{total} = Error_{TS} + Error_{aiding} + Error_{aspheric} \tag{1}$$

In this equation, $Error_{total}$ is the total error of the measurement system, $Error_{TS}$ is the error of the TS (transmission sphere, a standard lens used on the interferometer), and $Error_{aiding}$ is the error of the aided element (such as flat mirror or convex sphere mirror). $Error_{aspheric}$ is the error of the off-axis aspheric under testing. Usually, the quality of TS is so high that $Error_{TS}$ is very small. If the $Error_{aiding}$ is also much smaller than $Error_{aspheric}$, the test result $Error_{total}$ is approximately equal to $Error_{aspheric}$. Otherwise, when the $Error_{aiding}$ is not small enough, we can subtract $Error_{aiding}$ from $Error_{total}$ to obtain $Error_{aspheric}$ after calibrating the $Error_{aiding}$. However, for off-axis aspheric surfaces, it is difficult to adjust all the components to the correct location. If there exists alignment error, there will be unavoidable misalignment errors such as astigmatism and coma in the test results. As a result, the adjustment turns into the most difficult process within the auto-collimation method.

The limitation of the auto-collimation method is attributed to the surface type and aperture of the off-axis aspheric under testing. As the aperture of the aiding element should be larger than the off-axis aspheric, this is a difficult and expensive mission if the aperture is up to meter level.

2.2. Single Computer Generated Hologram (CGH) Method

As a diffractive optical element, CGH can produce any shape wavefronts [9]. Figure 2 shows the principle of this method. The interferometer, CGH, and off-axis aspheric under testing are all on the same axis, so the aberration that should be compensated will be reduced [16].

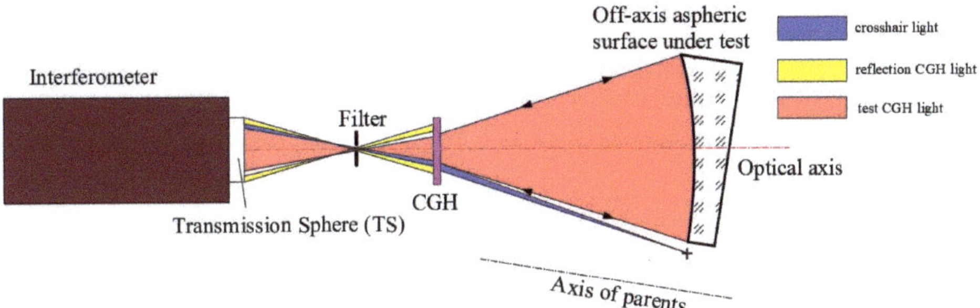

Figure 2. Schematic diagram of single computer generated hologram (CGH) method for off-axis aspheric surface.

In Figure 2, the output wavefront after TS is the convergent sphere wavefront. However, the CGH changes the sphere wavefront into the off-axis aspheric wavefront to match the theoretical shape of the off-axis aspheric surface under testing, where this off-axis aspherical wavefront vertically illuminates on the surface. After that, the wavefront with the information of the off-axis aspheric surface is reflected again into CGH via the same path it comes from, and ultimately back to the interferometer to form interference fringes, from which we eventually gain the test results. The test result with the single CGH method is also involved in several errors, as shown in Equation (1), by replacing $Error_{aiding}$ with $Error_{CGH}$. Before the CGH basement reaches the precision requirement, it should be processed to achieve a high quality in surface shape and parallelism, and then processed by photoetching to ensure the quality of the CGH [17]. As a result, the $Error_{CGH}$ is too small to have an effect on the testing results.

The CGH includes three parts: the test CGH, reflection CGH, and crosshair CGH. The test CGH is the transmission diffraction part (the red part in Figure 2), which creates the same wavefront of the off-axis aspheric surface under testing. When the off-axis aspheric surface is different, the corresponding CGH varies according to the customized design. Reflection CGH is used to align the interferometer and CGH (the yellow part in Figure 2), and crosshair CGH is applied to align the CGH and off-axis aspheric under testing (the blue part in Figure 2).

Within the single CGH method, the CGH design is the most important part. The geometry parameters in Figure 2 determine the size of the CGH and the aberration to be compensated. As it is a key parameter in the single CGH method, the aberration should be compensated by the CGH diffraction wavefront that decides the measurement accuracy and can be designed with Zemax (Version June 9, 2009). After obtaining the CGH phase function, the fringe etching position can be calculated with a special MATLAB program (R2016b). The design process is shown in Figure 3.

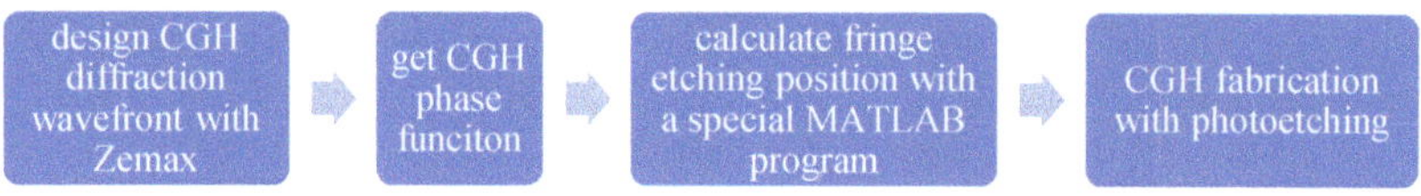

Figure 3. CGH design process.

The disadvantage of the single CGH method lies in the minimum fringe spacing and the size of the whole CGH. When the minimum fringe spacing is less than the ability of photoetching, this customized CGH cannot be fabricated successfully.

2.3. Hybrid Compensation Method

Hybrid compensation is also an interference measurement technique to solve the shape measurement problem of the off-axis aspheric surface. When the aberration is larger than what a single CGH cannot be compensated, hybrid compensation is a suitable choice. In Figure 4, a fold sphere mirror was used to compensate most of the low-order aberration and a CGH was applied to compensate the residual aberration, which is still a null test system [18]. The function of each element in Figure 4 is similar to that illustrated in Figure 2 of Section 2.2.

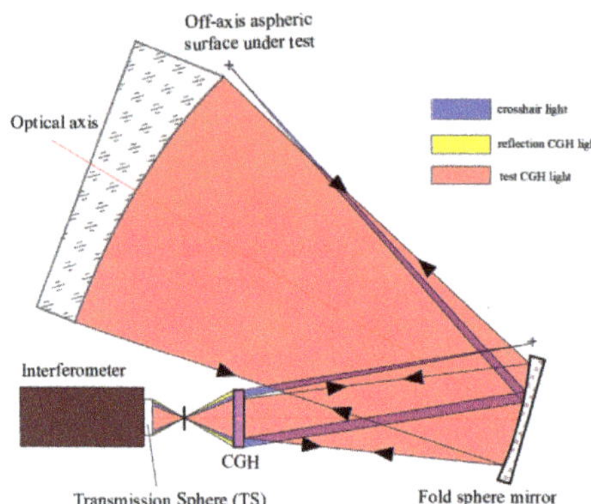

Figure 4. Schematic diagram of the hybrid compensation method for off-axis aspheric surface.

The test result of hybrid compensation includes four parts, as seen in Equation (2),

$$Error_{total} = Error_{TS} + Error_{CGH} + Error_{sphere} + Error_{aspheric} \tag{2}$$

In this equation, the definition of each sub-term is similar to that in Equation (1). The $Error_{CGH}$ from the CGH is very difficult to measure directly and can be estimated by an indirect method [19,20]. Similar to the first two methods (Sections 2.1 and 2.2), when $Error_{TS}$, $Error_{CGH}$, and $Error_{sphere}$ are all small enough, the test result $Error_{total}$ is approximately equal to $Error_{aspheric}$. Otherwise, the error of these elements should be separately calibrated and then subtracted from the test result to obtain the error of the off-axis aspheric under testing.

In this measurement system, the fold sphere and CGH should be designed. According to laboratory conditions, the fold sphere is chosen by selecting the main parameters of the curvature radius and aperture. Meanwhile, the location position and fold angle of this fold sphere mirror should consider whether it will disturb the measurement. After choosing the fold sphere mirror and determining the geometrical parameters of its measurement system, the CGH can be designed. The CGH design processing is the same as that of the single CGH method, which is shown in Figure 3.

As shown in Figure 4, this hybrid compensation measurement system possesses four elements of the folded optical axis, resulting in a complex system that is characterized with huge difficulty in adjusting. To reduce the difficulty, the customized CGH is divided into multiple parts. In addition to testing the CGH, others are used to align between different elements such as the interferometer and CGH, the fold sphere and CGH, and the aspheric and CGH.

The proposed three measurement methods that can all test the shape accuracy of conic off-axis aspheric surface belong to the null test. The auto-collimation method, which is the simplest technique with the simplest aiding element, is possessed of stronger generality. Single CGH and hybrid compensation both need customized CGH, so they have less generality. The customized CGH, which is a diffractive element, needs long and expensive preparation. However, these two methods can measure more surface types with the assistance of adjustment marks, resulting in a wider range of application.

3. Experiments

To verify the correctness of these methods, we used three methods to measure the same conic off-axis aspheric surface. According to the conditions of our laboratory, an off-axis paraboloid (OAP) was chosen and its parameters are shown in Table 1.

Table 1. The parameters of the chosen off-axis paraboloid (OAP).

Type of Aspheric	Aperture	Conic	Vertex Radius of Curvature	Off-Axis Distance
Off-axis paraboloid	135 mm	−1	1000 mm	165 mm

3.1. Auto-Collimation Method

First, we used the classical auto-collimation method to measure this OAP. The result can be seen as the real shape distribution of this OAP. According to Figure 1a, this OAP is located on the off-axis position. Meanwhile, a $\Phi150$ mm flat mirror was chosen as the aiding optical element. A 6-inch Zygo interferometer (Middlefield, CT, USA) with an F/0.8 transmission sphere (TS) was used. The quality of this TS was $1/10\lambda$ (PV), and the shape error of the flat mirror measured via interferometry was about 0.1λ (PV). The corresponding error distribution is seen in Figure 5. According to Equation (1), the testing results can be reckoned as the error of OAP because the errors of TS and flat mirror were both far less than that of the OAP.

Due to its location on an off-axis position in Figure 5, the OAP was adjusted to guarantee the position accuracy by a five-dimension adjusting tool. In this measurement system, the adjustment processing is very difficult because of a lack of alignment marks. Coma, astigmatism, and power aberration always exist because of the misalignment of different optical elements. Therefore, adjusting experience is critical. After repeating the adjusting until there is almost no coma and astigmatism, a credible testing result will be obtained.

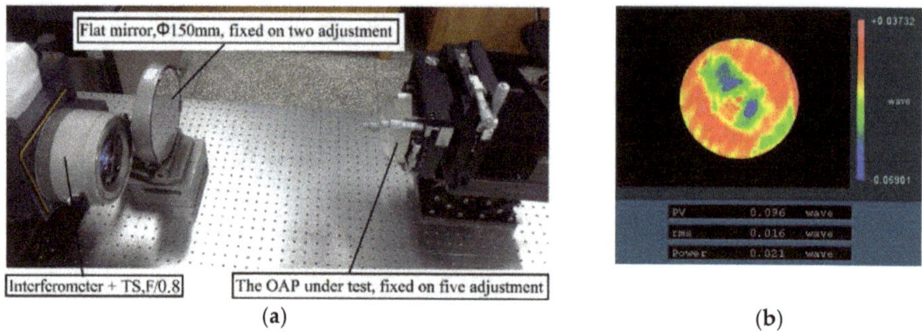

(**a**)   (**b**)

Figure 5. (**a**) Photo of auto-collimation for this OAP; (**b**) the error of flat mirror (PV = 0.096λ).

3.2. Single CGH Method

Second, we tested this OAP with the single CGH method. Depending on the conditions of our laboratory, silica (Φ58 mm, n = 1.457081 and d = 6.028 mm) was chosen as the CGH basement. Before processing the CGH, this silica basement was processed by IBF (ion beam figuring) to guarantee that its shape error was less than 5 nm (RMS) and its wedge angle was less than 10 s. The geometry parameters of this measurement system are defined in Figure 6, from which the designing result of a customized single CGH could be conducted and is demonstrated in Figure 7. The CGH wavefront was designed using Zemax software and the etching map was generated by the specialized MATLAB program (Figure 3).

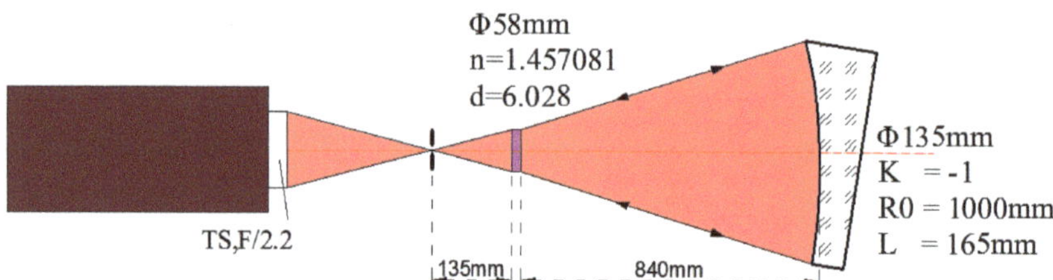

Figure 6. The geometry parameters of measuring this OAP with a single CGH.

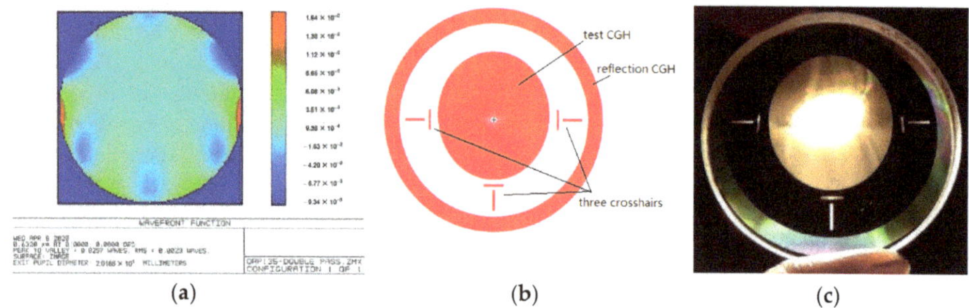

(**a**)   (**b**)   (**c**)

Figure 7. (**a**) Simulated residual wavefront of the single CGH method (test CGH) in Zemax (PV = 0.0257λ, RMS = 0.0023λ); (**b**) simulation pattern of the customized CGH; (**c**) CGH photograph.

After obtaining the design result of the customized CGH seen in Figure 7b, photoetching was used to fabricate the CGH with high manufacturing precision. Even so, CGH still had some errors and the real error of CGH was very difficult to calibrate directly. According to the corresponding estimation method of the CGH error, the total CGH error was about 7.8 nm (RMS), which is a negligible value compared with the error of the OAP [19,20]. The customized single CGH includes three parts: the test CGH (the inner part), reflection CGH (the outer part), and three crosshairs (three orthogonal rectangular couples), which are attributed to different functions. The test CGH, which is a transmission CGH using first order diffraction, was fabricated by a grating groove depth of 692.2 nm. It can be seen that the test CGH is the most important because it was adopted to measure the shape of the OAP. As shown in Figure 7a, the residual wavefront of test CGH was only 0.0257λ (PV) and 0.0023λ (RMS). With a grating groove depth of 158.2 nm, the reflecting CGH, which made use of the third order diffraction, was used to align the interferometer and CGH. With a grating groove depth of 692.2 nm, three crosshairs, which were transmission CGHs using first order diffraction, were applied to align the OAP and CGH. The combined CGH was fabricated by photoetching, as shown in Figure 7c. The measurement experiment is shown in Figure 8 with an obvious crosshair. Relying on these crosshairs, it was easy to adjust the various optical elements to its own correct position.

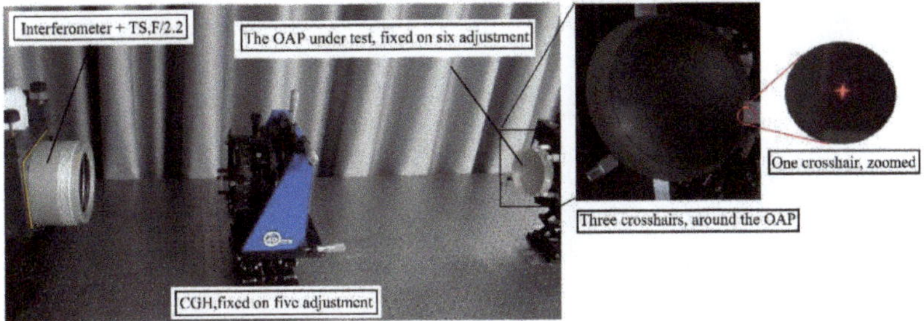

Figure 8. Photo of measuring the OAP with the single CGH method.

3.3. Hybrid Compensation Method

Third, we tested the OAP using the hybrid compensation method. In the same way, the CGH basement (silica) and the fold sphere mirror were chosen according to our laboratory, and the related geometry parameters are shown in Figure 9. Ultimately, the design result of the customized CGH for the hybrid compensation system is illustrated in Figure 10.

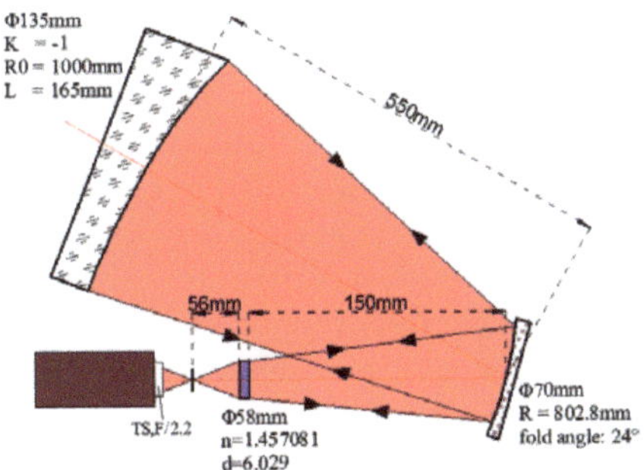

Figure 9. The geometry parameters of measuring this OAP via the hybrid compensation method.

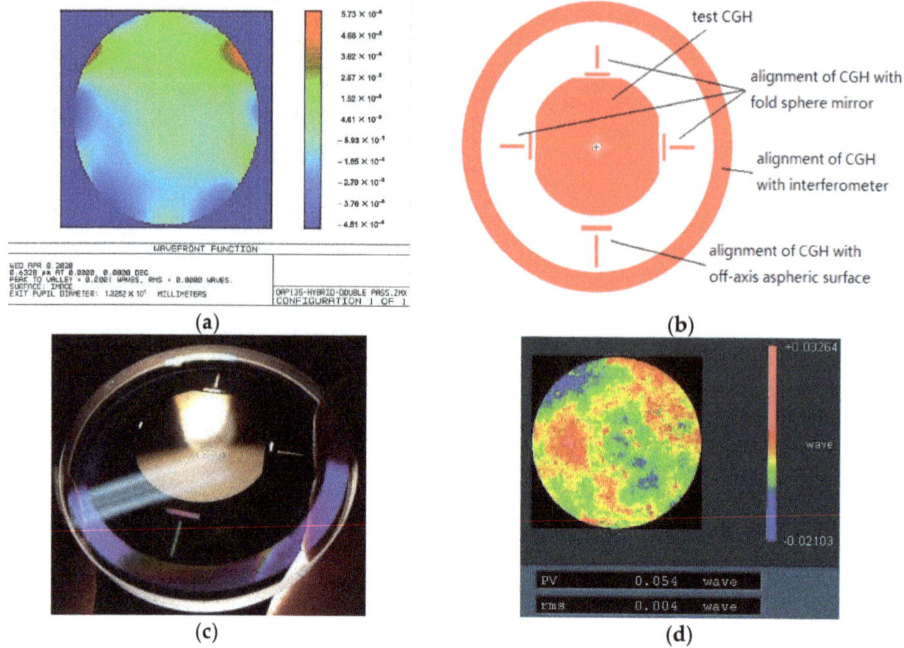

Figure 10. (**a**) Simulated residual wavefront of the hybrid compensation method (test CGH) in Zemax (PV = 0.0001λ); (**b**) simulation pattern of the customized CGH for this hybrid compensation system; (**c**) CGH photograph; (**d**) error of the fold sphere mirror (PV = 0.054λ).

Another multiple combined CGH was designed and produced. The accuracy of the CGH basement was the same as that of the basement in Section 3.2. As shown in Figure 10, the multiple combined CGH was involved in four parts such as the test CGH, alignment CGH and so on. Three parts were transmission CGH, except the part aligning CGH with the interferometer. The test CGH and alignment CGH with the fold sphere mirror both adopted the first order diffraction with the same grating groove depth of 692.2 nm. However, other parts adopted the third order diffraction with the corresponding groove depth of 158.2 nm. The residual wavefront of the test CGH was only 0.0001λ (PV), as shown in Figure 10a. At the same time, the error of the fold sphere mirror, which was measured via interferometry, was about 0.05λ (PV), far less than the error of this OAP.

During the measuring process, the customized CGH was placed close to the fold sphere mirror by a customized fixture, with which the relative position between the CGH and fold sphere could be guaranteed, as presented in Figure 11. At the same time, the customized fixture was mounted on a five-dimensional adjuster so that they would be regarded as the whole during the adjusting process. After repeatedly adjusting, the position error of each element can be reduced to a negligible value. Meanwhile, the errors of the fold sphere mirror and CGH were far less than the error of OAP, so the testing result can be regarded as the shape error distribution of the OAP.

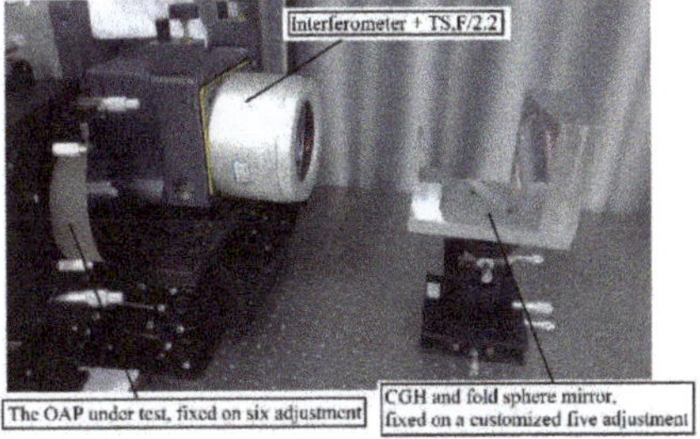

Figure 11. Photo of measuring this OAP via hybrid compensation (with customized fixture).

3.4. Experimental Results with These Three Methods

This OAP was measured using three methods, respectively. After the measured data were simply processed in MetroPro software (V9.1.1) [21] to remove piston, tilt, and power, the experimental results were obtained and are shown in Figure 12.

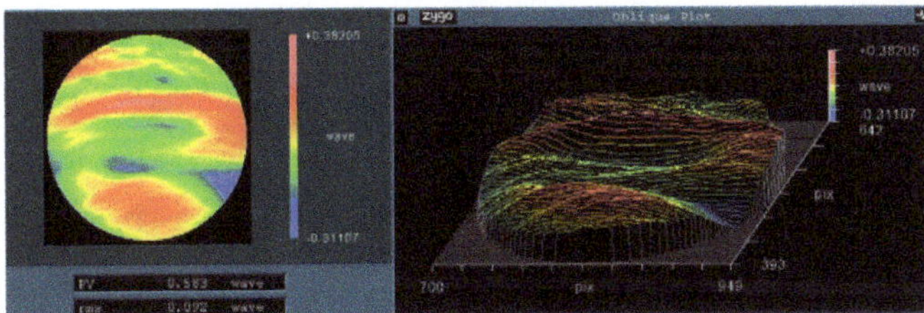

(**a**) Test result with the auto-collimation method

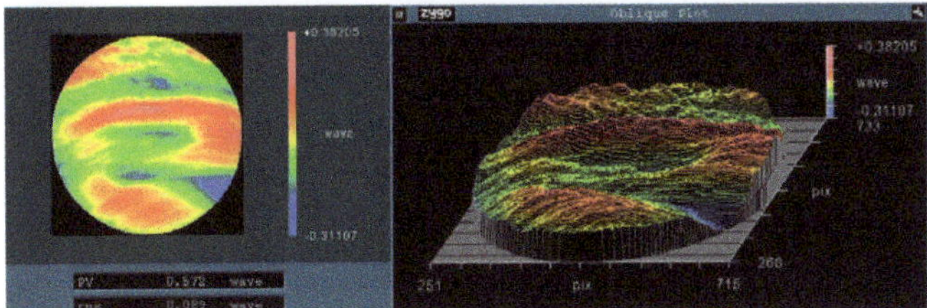

(**b**) Test result with the single CGH method

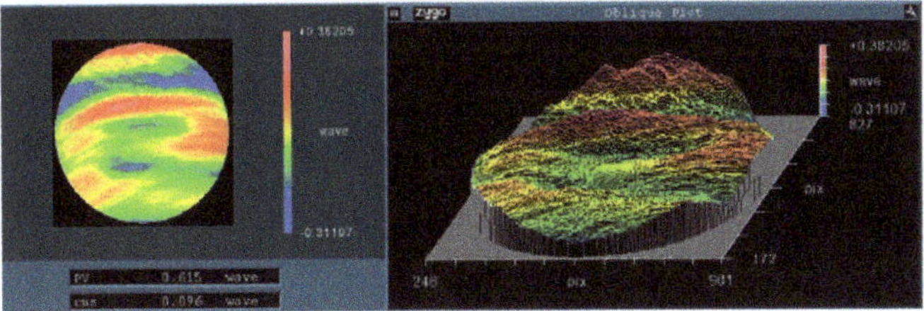

(**c**) Test result with the hybrid compensation method

Figure 12. The test result of Φ135 mm OAP: (**a**) auto-collimation; (**b**) single CGH; (**c**) hybrid compensation.

The resulting value of auto-collimation method as PV = 0.583λ and RMS = 0.092λ, the resulting value of the single CGH method as PV = 0.572λ and RMS = 0.089λ, and the resulting value of the hybrid compensation method was PV = 0.615λ and RMS = 0.096λ. The testing results with different methods were nearly the same, although there was a slight difference on the shape distribution, which was mainly caused by the mapping distortion of CGH. Furthermore, there were a few small residual adjustment errors, some of which were particularly induced by the angle adjustment in the hybrid compensation method, as it is a delicate process. Hence, a more accurate alignment technique is still being investigated.

4. Conclusions

Shape accuracy, as a key parameter of the off-axis conic aspheric surface, has an important effect on the imaging quality of the optical system, so it is necessary to guarantee the shape is accurate during the measuring process. In this study, we provided three methods to test the shape accuracy of the off-axis conic aspheric surface with high precision. The first was auto-collimation, which belongs to a classical and simple method with a requirement of a flat mirror or a sphere mirror as the aiding element. However, this method has some limitations to the surface type and the aperture of off-axis surface under testing. The second was the single CGH, which is widely used as an effective method with alignment marks to reduce the adjustment difficulty. However, this method requires a customized CGH, which varies with changes in off-axis aspheric surfaces, making it an expensive option. With the need of a fold sphere mirror and a customized CGH, the third is hybrid compensation, which belongs to a more complex measurement technology. The method has a stronger aberration compensation ability and is more suitable for the off-axis aspheric surface of large aperture and large asphericity. Unfortunately, its adjustment is more difficult. Therefore, we recommend not using this method unless absolutely necessary.

In this study, an OAP was measured via these three methods. By the means of auto-collimation, a $\Phi150$ mm flat mirror was used as the aiding element, and the result was PV = 0.583λ and RMS = 0.092λ; by the means of the single CGH, a customized CGH was designed and fabricated, and the result was PV = 0.572λ and RMS = 0.089λ; by the means of hybrid compensation, a fold sphere mirror was chosen, and another customized CGH was also designed and fabricated where the result was PV = 0.615λ and RMS = 0.096λ. These three measurement methods brought forth approximate results, in the meantime, the shape distributions were also close to each other, which proves that these three measurement methods can all obtain comparatively accurate testing results. Furthermore, these three methods can also cross-check the correctness of each other and other available methods.

Author Contributions: Methodology, S.L.; software, Y.X.; investigation, X.L.; resources, H.L.; data curation, J.Z.; writing of the original draft preparation, S.L.; writing of review and editing, W.L., H.L. and X.L.; visualization, Y.X.; supervision, J.Z. and W.L.; project administration, W.L. and H.L. All authors have read and agreed to the published version of the manuscript.

Funding: This work was supported by the National Natural Science Foundation of China (No. 61505157), the Xi'an Science and Technology Board (No. 201805031YD9CG15(2)), the Science and Technology Program of Shaanxi Province (No. 2020GY-045), and the Xi'an Key Laboratory of Intelligent Detection and Perception (No. 201805061ZD12CG45).

References

1. Yan, F.; Tao, X. Experimental study of an off-axis three mirror anastigmatic system with wavefront coding technology. *Appl. Opt.* **2012**, *51*, 1749–1756. [CrossRef] [PubMed]
2. Kim, Y.-S.; Ahn, K.-B.; Park, K.-J.; Moon, I.-K.; Yang, H.-S. Accuracy Assessment for Measuring Surface Figures of Large Aspheric Mirrors. *J. Opt. Soc. Korea* **2009**, *13*, 178–183. [CrossRef]
3. Meinel, A.B. Optical Testing of Off-Axis Parabolic Segments without Auxiliary Optical Elements. *Opt. Eng.* **1989**, *28*, 280171. [CrossRef]

4. Wang, X.-K.; Zheng, L.-G.; Zhang, B.-Z.; Li, R.-G.; Fan, D.; Deng, W.-J.; Wang, X.; Zhang, F.; Zhang, Z.-Y.; Zhang, X.-J. Test of an off-axis asphere by subaperture stitching interferometry. In Proceedings of the 4th International Symposium on Advanced Optical Manufacturing and Testing Technologies: Optical Test and Measurement Technology and Equipmen, Chengdu, China, 20 May 2009; Society of Photo Optical Instrumentation Engineers (SPIE): Bellinghan, WA, USA, 2009; Volume 72832. [CrossRef]

5. Wen, Y.; Cheng, H. Measurement for off-axis aspheric mirror using off-axis annular subaperture stitching interferometry: Theory and applications. *Opt. Eng.* **2015**, *54*, 14103. [CrossRef]

6. Zhao, Z.; Zhao, H.; Gu, F.; Du, H.; Li, K. Non-null testing for aspheric surfaces using elliptical sub-aperture stitching technique. *Opt. Express* **2014**, *22*, 5512–5521. [CrossRef] [PubMed]

7. Burke, J.; Wang, K.; Bramble, A. Null test of an off-axis parabolic mirror I Configuration with spherical reference wave and flat return surface. *Opt. Express* **2009**, *17*, 3196. [CrossRef] [PubMed]

8. Ahn, K.-B.; Kim, Y.-S.; Lee, S.; Park, K.; Kyeong, J.; Park, B.-G.; Park, C.; Lyo, A.-R.; Yuk, I.-S.; Chun, M.-Y. Sensitivity analysis of test methods for aspheric off-axis mirrors. *Adv. Space Res.* **2011**, *47*, 1905–1911. [CrossRef]

9. MacGovern, A.J.; Wyant, J.C. Computer Generated Holograms for Testing Optical Elements. *Appl. Opt.* **1971**, *10*, 619. [CrossRef] [PubMed]

10. Kim, T.; Burge, J.H.; Lee, Y.; Kim, S. Null test for a highly paraboloidal mirror. *Appl. Opt.* **2004**, *43*, 3614–3618. [CrossRef] [PubMed]

11. Wang, M.; Asselin, D.; Topart, P.; Gauvin, J.; Berlioz, P.; Harnisch, B. CGH Null-test Design and Fabrication for Off-axis Aspherical Mirror Tests. *Int. Opt. Des.* **2006**, *6342*, 63421. [CrossRef]

12. Talha, M.; Chang, J.; Wang, Y.; Cheng, D.; Zhang, T.; Aslam, M.; Khan, A.N. Computer generated hologram null compensator for optical metrology of a freeform surface. *Optik* **2010**, *121*, 2262–2265. [CrossRef]

13. Li, F.Z.; Luo, X.; Zhao, J.L.; Xue, D.L.; Zheng, L.G.; Zhang, X.J. Test of off-axis aspheric surfaces with CGH. *Opt. Precis. Eng.* **2011**, *19*, 709–716. [CrossRef]

14. Burge, J.H.; Kot, L.B.; Martin, H.M.; Zehnder, R.; Zhao, C. Design and analysis for interferometric measurements of the GMT primary mirror segments. *SPIE Astron. Telesc. Instrum.* **2006**, *6273*, 62730. [CrossRef]

15. Oh, C.J.; Lowman, A.E.; Smith, G.A.; Su, P.; Huang, R.; Su, T.; Kim, D.; Zhao, C.; Zhou, P. Fabrication and testing of 4.2m off-axis aspheric primary mirror of Daniel K. Inouye Solar Telescope. *Adv. Opt. Mech. Technol. Telesc. Instrum. II* **2016**, *9912*, 99120. [CrossRef]

16. Li, S.; Liu, B.; Tian, A.; Guo, Z.; Yang, P.; Zhang, J. A practical method for determining the accuracy of computer-generated holograms for off-axis aspheric surfaces. *Opt. Lasers Eng.* **2016**, *77*, 154–161. [CrossRef]

17. Xie, Y.; Chen, Q.; Wu, F.; Qiu, C.K.; Hou, X.; Zhang, J. Design and fabrication of twin computer-generated holograms for testing concave aspherical surfaces. *Opto-Electron. Eng.* **2008**, *35*, 40–43.

18. Li, S.; Zhang, J.; Liu, W.; Guo, Z.; Li, H.; Yang, Z.; Liu, B.; Tian, A.; Li, X. Measurement investigation of an off-axis aspheric surface via a hybrid compensation method. *Appl. Opt.* **2018**, *57*, 8220–8227. [CrossRef] [PubMed]

19. Chang, Y.-C.; Zhou, P.; Burge, J.H. Analysis of phase sensitity for binary computer-generated holograms. *Appl. Opt.* **2006**, *45*, 4223–4234. [CrossRef]

20. Zhou, P.; Burge, J.H. Fabrication error analysis and experimental demonstration for computer-generated holograms. *Appl. Opt.* **2007**, *46*, 657–663. [CrossRef] [PubMed]

21. Zygo Corporation. *MetroPro Manual Version 9.1.1*; Zygo Corporation: Middlefield, CT, USA, 2012.

Article

Investigation on the Adsorption-Interaction Mechanism of Pb(II) at Surface of Silk Fibroin Protein-Derived Hybrid Nanoflower Adsorbent

Xiang Li [1], Yan Xiong [1,*], Ming Duan [1,*], Haiqin Wan [2], Jun Li [1], Can Zhang [1], Sha Qin [1], Shenwen Fang [1] and Run Zhang [3]

[1] School of Chemistry and Chemical Engineering, Southwest Petroleum University, Chengdu 610500, China; 201721000245@stu.swpu.edu.cn (X.L.); 201821000253@stu.swpu.edu.cn (J.L.); 201922000213@stu.swpu.edu.cn (C.Z.); 201921000230@stu.swpu.edu.cn (S.Q.); mduana124@swpu.edu.cn (S.F.)

[2] State Key Laboratory of Pollution Control and Resource Reuse, Jiangsu Key Laboratory of Vehicle Emissions Control, School of the Environment, Nanjing University, Nanjing 210023, China; wanhq@nju.edu.cn

[3] Australian Institute for Bioengineering and Nanotechnology, AIBN, The University of Queensland, St Lucia, QLD 4072, Australia; r.zhang@uq.edu.au

* Correspondence: 201031010013@swpu.edu.cn (Y.X.); mduan@swpu.edu.cn (M.D.)

Received: 18 February 2020; Accepted: 6 March 2020; Published: 9 March 2020

Abstract: For further the understanding of the adsorption mechanism of heavy metal ions on the surface of protein-inorganic hybrid nanoflowers, a novel protein-derived hybrid nanoflower was prepared to investigate the adsorption behavior and reveal the function of organic and inorganic parts on the surface of nanoflowers in the adsorption process in this study. Silk fibroin (SF)-derived and copper-based protein-inorganic hybrid nanoflowers of SF@Cu-NFs were prepared through self-assembly. The product was characterized and applied to adsorption of heavy metal ion of Pb(II). With Chinese peony flower-like morphology, the prepared SF@Cu-NFs showed ordered three-dimensional structure and exhibited excellent efficiency for Pb(II) removal. On one hand, the adsorption performance of SF@Cu-HNFs for Pb(II) removal was evaluated through systematical thermodynamic and adsorption kinetics investigation. The good fittings of Langmuir and pseudo-second-order models indicated the monolayer adsorption and high capacity of about 2000 mg g^{-1} of Pb(II) on SF@Cu-NFs. Meanwhile, the negative values of $\Delta_r G^\theta_{m(T)}$ and $\Delta_r H^\theta_m$ proved the spontaneous and exothermic process of Pb(II) adsorption. On the other hand, the adsorption mechanism of SF@Cu-HNFs for Pb(II) removal was revealed with respect to its individual organic and inorganic component. Organic SF protein was designated as responsible 'stamen' adsorption site for fast adsorption of Pb(II), which was originated from multiple coordinative interaction by numerous amide groups; inorganic $Cu_3(PO_4)_2$ crystal was designated as responsible 'petal' adsorption site for slow adsorption of Pb(II), which was restricted from weak coordinative interaction by strong ion bond of Cu(II). With only about 10% weight content, SF protein was proven to play a key factor for SF@Cu-HNFs formation and have a significant effect on Pb(II) treatment. By fabricating SF@Cu-HNFs hybrid nanoflowers derived from SF protein, this work not only successfully provides insights on its adsorption performance and interaction mechanism for Pb(II) removal, but also provides a new idea for the preparation of adsorption materials for heavy metal ions in environmental sewage in the future.

Keywords: silk fibroin; hybrid nanoflowers surface; Pb(II) removal; interaction mechanism

1. Introduction

With fast growing activities of urbanization and industrialization, heavy metal ion (HMI) contamination in water environments has been widely brought by the rapid economic development [1]. Due to their rapid accumulation in the food chain and non-biodegradable properties, HMIs are regarded as one of the most serious contamination sources with highly toxicity and carcinogenicity even at trace amount exposure [2]. Pb(II) is an often encountered HMI which has been widely used in industries of batteries manufacturing, shipbuilding, oil mining, etc. [3]. The large amount of Pb(II) discharge in water environment and Pb(II) accumulation in human body can lead to physical defects such as nephropathy, hepatopathy, and encephalopathy [4,5]. Even more, high concentration of lead ions will do harm to children's health [6]. According to the guidelines set by WHO and EPA, the permissible limit of Pb(II) in portable water should not exceed 0.05 mg L^{-1} [7,8]. On account of the serious threatening on the ecosystem's sustainable development and human health, the removal of Pb(II) from waste water has become an urgent problem and a mandatory task for environmental protection [9].

Various treatment techniques—such as chemical reduction [10], biological conversion [11], membrane separation [12,13], and adsorption treatment [14]—have been developed and applied to remove HMIs during the past decades. With obvious advantages of high efficiency, cost-effectiveness and simple operation, adsorption technology has been regarded as one of the most effective and competitive methods for HMIs treatment [15–18]. Consequent, the development of functional adsorbent material and the application to efficient HMIs removal are highly desirable for water pollution treatment.

So far, a great number of materials—including lignin [19], biochar [20], chitosan [21], fabrics [22], soil [23], metal-organic frameworks (MOFs) [24], graphene oxide (GO) [25], and nanomaterials (such as nanofiber, nanobubble, and nanotube) [26–28]—have been studied and prepared as adsorbents for HMIs removal. Among these materials, organic–inorganic hybrid nanoflowers (HNFs) is newly developed functional material and has received considerable attention due to its distinctive physiochemical characteristics. By binding inorganic nanoparticles to organic components, HNFs show properties of simple product synthesis and high biomolecule efficiency comparing with the pure organic nanoflowers [29] and inorganic nanoflowers [30]. Since Ge et al. [31] first reported the preparation of BSA-incorporated Cu$_3$(PO$_4$)$_2$ nanoflowers, biomaterials-based HNFs have attracted increasing interest and many researches have focused on their biochemical applications of biosensing [32–34], biocatalysis [35–37], and drug delivery [38].

The organic component and the preparation method were two important aspects which would have great influence on the structure, morphology and property of the HNFs composites. For organic component of HNFs, protein is usually selected as a typical biological material for HNFs fabrication owing to its unique chemical structure and special biological property. A series of proteins, including serum albumin (BSA) [31], glucose oxidase (GOx), horseradish peroxidase (HRP) [39], and immunoglobulin G (Ig G) [40] have been employed to prepare HNFs. Although these nanoflowers show excellent performances, the products generally suffer from the disadvantages of high price and difficult acquisition of protein, which greatly limits the HNFs products in actual applications.

Silk fibroin (SF), a facile and low-cost protein which is obtained from the silkworm, is a well-known and widely-used natural macromolecular protein. During the past thousands of years, SF has been considered as an excellent raw material for the traditional use in textile industries [41]. Nowadays, the attractive properties of SF protein—such as good mechanism stability [42], superior biocompatibility [43], and excellent optic performances [44]—have made SF effective use in bioelectronic substrate [45], optical sensor [46], drug delivery [47,48], and so on. Consequently, SF protein has been regarded as an excellent candidate of organic biomolecules for HNF fabrication [49].

Besides the organic component, the preparation method is also of great importance for the HNFs preparation. If the biomolecules are improperly bonded or immobilized with the organic component, the prepared HNFs usually exhibit lower biomolecule activity, enhanced biomolecule mass-transfer limitations, and unfavorable conformational changes in the biomolecules [50]. Compared with conventional immobilization methods (such as covalent bond [51], physical trap [52]) and new fabrication techniques (such as welding [53], nanoimprinting [54]), self-assembly process have proven to show characteristics of simple synthesis, high efficiency, and bright prospect of enhancing stability, activity, and even selectivity of biomolecules for HNFs fabrication [55].

In this work, copper–protein hybrid nanoflowers by employing SF protein as natural biomaterial and copper phosphate as inorganic component are fabricated for efficient Pb(II) treatment by self-assembly method. The prepared nanoflowers derived from SF protein, denoted as SF@Cu-HNFs thereafter, exhibit several significant advantages: (1) raw biomaterials of SF protein are easy and cheap to obtain; (2) acidic amino acids in the primary structure of SF can bind cations to drive self-assembly easily; (3) abundant functional hydroxyl and amino groups are provided by SF protein for Pb(II) adsorption. The synthesized SF-based nanoflowers were characterized and applied to HMI adsorption (Pb(II), Ni(II), and Cd(II)). Compared with the adsorption performances of Cd(II) and Ni(II), the prepared SF@Cu-HNFs exhibited excellent adsorption selectivity and significant adsorption capacity for Pb(II) removal. Subsequently, the adsorption performance of SF@Cu-HNFs was systematically evaluated for Pb(II) adsorption through thermodynamic (adsorption isotherm and adsorption capacity) and adsorption kinetics investigation. Furthermore, the interaction mechanism of SF@Cu-HNFs was successfully revealed and verified for Pb(II) adsorption with respect to its individual component of organic SF protein and inorganic $Cu_3(PO_4)_2$ crystal.

2. Experimental

2.1. Reagents and Materials

Silk fibroin (SF) protein was purchased from Xi'an Shennong Biotechnology Co., Ltd. (Xi'an, China). The protein has been purified and used as it received. Cadmium nitrate $(Cd(NO_3)_2)$ was purchased from Aladdin Reagent Co., Ltd. (Shanghai, China). Lead nitrate $(Pb(NO_3)_2)$, nickel nitrate $(Ni(NO_3)_2)$, copper sulphate $(CuSO_4)$, sodium chloride (NaCl), potassium chloride (KCl), potassium dihydrogen phosphate (KH_2PO_4), dibasic sodium phosphate (Na_2HPO_4), sodium hydroxide (NaOH), and nitric acid (HNO_3) were obtained from Kelong Chemical Reagent Company (Chengdu, China). All the reagents were of analytical grade and used as received. Wahaha® purified water (Wahaha, Hangzhou, China) was used for the preparation of solutions and throughout the experiments.

SF solution was prepared in water and the concentration was adjusted to the value as needed by experiments. $Pb(NO_3)_2$, $Cd(NO_3)_2$, and $Ni(NO_3)_2$ stock solutions of 1×10^3 mg L^{-1} were prepared in water and working solutions were prepared freshly for daily use. The pH was adjusted by using small amounts of 0.1 mol L^{-1} NaOH or 0.1 mol L^{-1} HNO_3 solutions without significantly altering the HMIs concentration. Solution pH was monitored using a pH meter (PHS-3C, Yidian Inc., Ltd., Shanghai, China).

2.2. Synthesis of SF@Cu-HNFs via Self-Assembly

0.01 M phosphate buffer solution (PBS, pH = 7.4): Weigh 0.135 g of potassium dihydrogen phosphate, 0.71 g of disodium hydrogen phosphate, 4 g of sodium chloride, and 0.1 g of potassium chloride in order using an analytical balance, and add an appropriate amount purified water was stirred to dissolve it. The solution was transferred to a 500 mL volumetric flask, and the volume was adjusted with purified water. It was then transferred to a reagent bottle and refrigerate at 4 °C until use.

By utilizing SF protein as natural biomaterial and $Cu_3(PO_4)_2$ as inorganic component, the hybrid nanoflowers of SF@Cu-HNFs were synthesized according to similar methods developed for BSA-based and laccase-based nanoflowers described with some modification [31,56]. In brief, 4 mL of PBS (pH = 7.4) containing different SF concentration was firstly added with 40 μL $CuSO_4$ solution (100 mM). Then resultant mixtures were gently shaken for 5 min and followed by incubation at 25 °C with different preparation time. Finally, blue hybrid nanoflowers were collected, washed, with deionized water several times and dried by vacuum freeze-drying.

2.3. Characterization of SF@Cu-HNFs

The morphologies of the prepared SF@Cu-HNFs products were characterized by a scanning electron microscope (SEM, EV0 MA15, Carl Zeiss, Germany) at an acceleration voltage of 20 kV. All samples were sputter-coated with gold using an E1045 Pt-coater (Carl Zeiss, Germany) before SEM observation. Elemental analysis was conducted with an energy dispersive X-ray spectrometer (EDS) equipped in the SEM.

The crystal structures of the nanoflower products were characterized by X-ray diffraction (XRD) analysis (X Pert PRO MPD, PANalytical, Holland). Radial scans using Cu Kα radiation source at 20 mA and 40 kV were recorded in the reflection scanning mode from 2θ = 10 to 80° at a scanning rate of 1° min^{-1}.

The chemical structures of the nanoflower products were measured by Fourier transform infrared spectroscopy (Nicolet 6700 FTIR, Thermo Fisher Scientific Corp., USA) in the range of 400–4000 cm^{-1} with KBr pellets. The thermogravimetric analysis (TGA) of nanoflowers was measured with a thermogravimetric analyzer (STA449F3, Netzsch, Germany) in a dynamic atmosphere of dinitrogen with 20 cm^3 min^{-1} flow rate. The TGA measurements were performed with a temperature ranging from 40 to 700 °C in an alumina crucible at a rate of 5 °C min^{-1}.

To evaluate the surface adsorption and interaction process of Pb(II) on nanoflowers adsorbent, the surface characteristics of SF@Cu-HNFs were furthermore investigated with the addition of different Pb(II) concentrations. The surface zeta potential was determined by dynamic light scattering (DLS) measurements (NANO ZS, Malvern Instruments Ltd., UK) equipped with the DTS Ver. 4.10 software package.

2.4. Adsorption Performances of SF@Cu-NFs

Under the optimized pH condition, the adsorption performances of SF@Cu-NFs were studied by thermodynamic, adsorption kinetics, and selective experiments for Pb(II) removal. The Pb(II) concentration was detected through an atomic absorption spectrophotometer (AA-7020, Beijing East West Analysis Instrument Co., Ltd., Beijing, China) during the whole experiment.

2.4.1. Adsorption Kinetics Experiment

For the method, 30 mg of SF@Cu-NFs was first dispersed into 300 mL 300 mg L^{-1} Pb(II) solution under continuous stirring. The suspension was sealed and oscillated at room temperature to ensure equilibration. Then 3 mL of the suspension sample was taken from the system for filtration at a regular interval time. The residual Pb(II) concentration in the solution was also detected by AAS measurement like for thermodynamic adsorption. The amount of Pb(II) adsorbed on SF@Cu-NFs (Q_t) was calculated by subtracting the concentration of free Pb(II) at the time of t from the initial Pb(II) concentration as

$$Q_t = \frac{(C_0 - C_t) \times V}{M} \tag{1}$$

where C_0 and C_t (mg L^{-1}) were the initial and t time Pb(II) concentrations in liquid-phase, V (L) was the taken volume of dye solution, and M (g) was the mass of the SF@Cu-NFs adsorbent used. The data obtained were used to draw the kinetic adsorption curves for pseudo-first-order, pseudo-second-order, and intraparticle diffusion analysis.

2.4.2. Adsorption Isotherm Experiment and Adsorption Thermodynamics

The thermodynamic adsorption experiment of SF@Cu-NFs was carried out by a typical batch method. First, 7 pieces of 1 mg washed and dried SF@Cu-NFs were added into 7 pieces of 10.0 mL different concentration Pb(II) solutions (5, 20, 50, 80, 100, 400, 500 mg L^{-1}) placed in the tube at 298 K, respectively. Then the suspensions were sealed and were vibrated for 2 h at room temperature to ensure the complete adsorption. After filtrating the mixture for solid–liquid separation, the residual

Pb(II) concentration in the solution was detected by AAS measurement. The amount of Pb(II) adsorbed on SF@Cu-NFs (Q_e) was calculated by subtracting the concentration of Pb(II) from the initial concentration as

$$Q_e = \frac{(C_0 - C_e) \times V}{M} \tag{2}$$

where C_0 and C_e (mg L^{-1}) were the initial and equilibrium Pb(II) concentrations in liquid-phase, V (L) was the volume of Pb(II) solution, and M (g) was the mass of the SF@Cu-NFs adsorbent used. The data obtained were used to draw the adsorption isotherms for Langmuir, Freundlich, and Temkin analysis.

In order to obtain the experimental parameters of adsorption thermodynamics on the Pb(II) adsorption by SF@Cu-NFs, the adsorption capacity at three different temperatures of 298 K, 308 K, and 328 K were investigated with the initial Pb(II) concentration of 100 and 500 mg L^{-1}, respectively.

3. Results and Discussion

3.1. Condition Optimization for SF@Cu-NFs Preparation

In order to obtain the best formation morphology of SF@Cu-HNFs product, preparation conditions of SF concentration and reaction time were systematically investigated and optimized.

3.1.1. Effect of SF Protein on SF@Cu-NFs Formation

In order to investigate the effect of SF protein on the nanoflower formation, products with and without SF were firstly prepared and characterized. For the preparation of SF@Cu-NFs in this work, SF protein was simply added into the solution containing copper ions and phosphate [57]. Meanwhile, Cu$_3$(PO$_4$)$_2$ particles were also prepared by the similar procedures without the addition of SF solution. With 24 h preparation time, the photography and SEM images were shown in Figure 1a for SF@Cu-NFs products prepared with 400 mg L^{-1} SF and were shown Figure 1b for Cu$_3$(PO$_4$)$_2$ particles without SF, respectively. As it can be seen, although the appearances were similar (insets in Figure 1a,c), the microscopic images showed significant differences between SF@Cu-NFs products and Cu$_3$(PO$_4$)$_2$ particles. As shown in Figure 1a,b, SF@Cu-NFs products were successfully formed by self-assembly with uniform flower-like morphology similar to the Chinese national flowers peony (Inset in Figure 1b). However, Cu$_3$(PO$_4$)$_2$ particles with irregular flakes and uneven particle size distribution were formed without SF addition (as shown in Figure 1c,d). The Cu$_3$(PO$_4$)$_2$ particles were not flower-like but Chinese tremella like with loose structures (Inset in Figure 1d). The above results confirmed that SF protein was a key factor and had beneficial effects for the flower-like nanoflowers formation.

3.1.2. Effect of SF Concentration on SF@Cu-NFs Formation

Under the condition of 12 h reaction time, the effect of SF concentration on the product formation was investigated ranging from 50 to 400 mg L^{-1}. As shown in Figure 2, it was notable to observe that there was a great variation in morphology of SF@Cu-NFs by regulating the protein concentrations. Interestingly, the SF@Cu-NFs became smaller with the increase of SF concentration (Figure 2a,b,e,f), which may be caused by the increasing number of nucleation sites on the SF molecular. However, some SF@Cu-NFs products would bind with each other with SF concentration at 200 and 400 mg L^{-1} (Figure 2c,d,g,h), which was also disadvantageous for the adsorption with decreasing the SF@Cu-NFs number and surface. As a result, 100 mg L^{-1} SF concentration was selected to prepare the SF@Cu-NFs with the proper product size and superior shape.

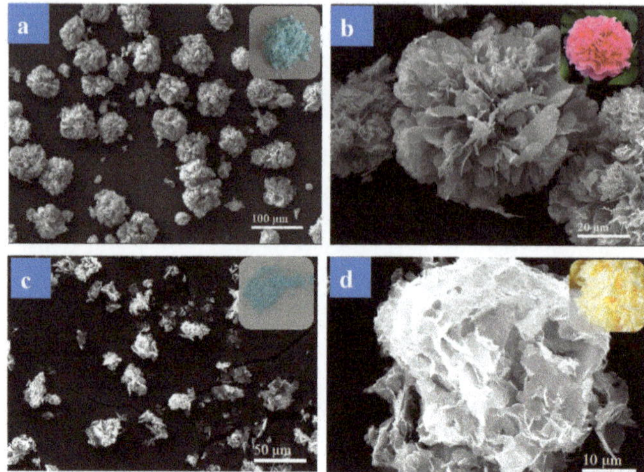

Figure 1. (**a,b**) SEM images of SF@Cu-NFs product with SF. Insets are photographies of SF@Cu-NFs and Chinese penoy flower. (**c**) and (**d**) SEM images of $Cu_3(PO_4)_2$ particles without SF. Insets are photographies of $Cu_3(PO_4)_2$ particles and Chinese tremella.

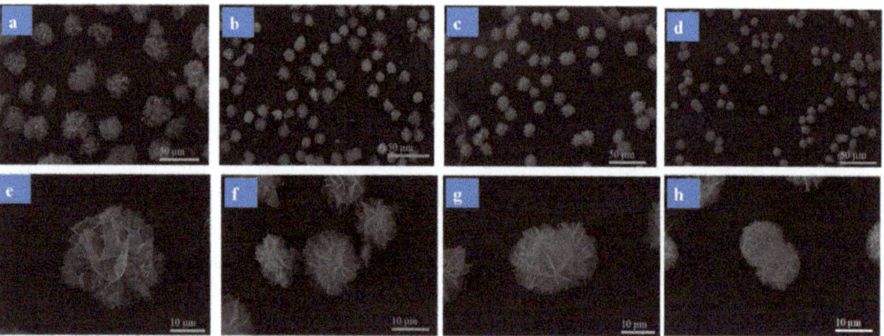

Figure 2. (**a–d**) SEM images of SF@Cu-NFs with different SF concentrations at 50 mg L^{-1}, 100 mg L^{-1}, 200 mg L^{-1} and 400 mg L^{-1}, respectively. (**e–h**) The orresponding SEM images enlarged with mangnification of individual products.

3.1.3. Effect of Reaction Time on SF@Cu-NFs Formation

In order to study the formation process of the three-dimensional hierarchical structures, the effect of preparation time on the product formation was investigated with the addition of 100 mg L^{-1} SF concentration. Experiments were carried out by collecting samples from the reaction mixture and observing intermediates and products at different time intervals. As shown in Figure S1, the PBS solution containing SF changed into blue after the addition of $CuSO_4$. Then the solution became turbid blue after 10 min, indicating the production of NFs products. The SEM images in Figure 3a–d and insets showed the appearance and solution changes of SF@Cu-NFs product ranging from 30 min to 24 h. The diameter distribution of products from 0 to 24 h was shown in Figure S2. The process of product formation can be divided into four corresponding stages.

The first stage is initial stage with reaction time from 10 to 30 min. At this stage, blue fine and visible particles began to appear in the solution (shown as inset in Figure 3a). As observing from corresponding SEM shown in Figure 3a, primary crystal of $Cu_3(PO_4)_2$ was formed (encircled in red circle) and SF protein molecules complexed with Cu^{2+} on its surface (encircled in green circle).

The product composited mainly through the coordination of amide groups in the protein backbone and was beneficial for the formation of larger nanosheet petals.

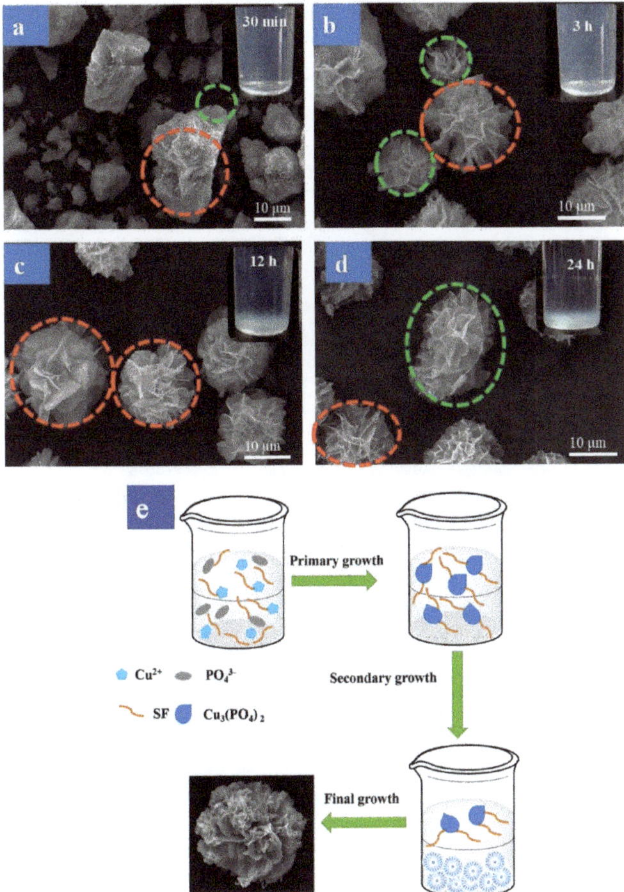

Figure 3. (**a–d**) SEM images of the nanostructures of SF@Cu-NFs products at different preparation times of 30 min, 3 h, 12 h, and 24 h. Insets are photographies of solution change. (**e**) Schematic illustration of the formation process of SF@Cu-NFs products at preparation time of 12 h.

The second stage is growth stage with reaction time from 30 min to 6 h. At this stage, the blue particles grew bigger and blue flocculent precipitation was observed in the solution (shown as inset in Figure 3b). As observing from corresponding SEM shown in Figure 3b, a series of SF@Cu-NFs products with complete flower-like shape have been formed (encircled in red circle). However, the morphology SF@Cu-NFs products were not uniformed and there were still some small petal products (encircled in green circle). This result indicated the nanoflowers needed to grow further.

The third stage is the formation stage with reaction time from 6 h to 12h. At this stage, small product particles furthermore grew up and deposited to the bottom ((shown as inset in Figure 3c)). As observing from corresponding SEM shown in Figure 3c, SF@Cu-NFs products with uniform flower-like shape and size were formed.

The fourth stage is the overgrowth stage with reaction time from 12 h to 24 h. At this stage, some SF@Cu-NFs grew up and deposited to the bottom (shown as inset in Figure 3d). This was because polar side chain of SF could promote the formation of large folding nanosheets through its hydroxyl, carboxyl, and amino groups. As observing from corresponding SEM shown in Figure 3d, the further growth made some SF@Cu-NFs bind with each other to form much bigger product (encircled in green circle). The overgrowth effect decreased the SF@Cu-NFs number and surface, which was disadvantageous for the adsorption. As a result, the incubation time of 12 h was considered to be the optimum preparation time for the SF@Cu-NFs products. The schematic illustration of the formation process of SF@Cu-NFs products at preparation time of 12 h was shown in Figure 3e.

3.2. Characterization of SF@Cu-NFs Product

3.2.1. Surface Morphology Measurement by LSCM and SEM Image

The morphologies of the synthesized SF@Cu-NFs were observed using and SEM images (shown in Figure 4a,b). As seen from SEM images, the prepared SF@Cu-NFs displayed highly peony flower-like morphology with diameters of about 50 μm.

Figure 4. (**a**,**b**) SEM images of SF@Cu-NFs product with low and high magnification; (**c**) EDS pattern of SF@Cu-NFs product.

Meanwhile, the EDS result was shown in Figure 4c and the related EDS data were listed in Table S1. The results identified the chemical species of the SF@Cu-NFs products and confirmed the presence of Cu, P, C, and O. The C and O can be attributed to SF protein and Cu, P and O can be attributed to $Cu_3(PO_4)_2$. Meanwhile, the appearance of Cl and Na may be brought by the residual PBS. However, the N element was not detected in the prepared SF@Cu-NFs product by EDS analysis. Because SF protein was generally considered as nitrogen rich [58], the abnormal absence of N may be attributed to the abundant O and C in the SF@Cu-NFs product, which would cover up the N peak in EDS.

3.2.2. Chemical Structure Investigation by FTIR and XRD

The phase structures of the as-prepared $Cu_3(PO_4)_2$ and SF@Cu-NFs were investigated by the XRD analysis and shown in Figure 5a. As observed, the diffraction peaks of SF@Cu-NFs and unmodified $Cu_3(PO_4)_2$ were in good agreement with the Joint Committee on Powder Diffraction File data for $Cu_3(PO_4)_2$ (File NO. 00-022-0548). As a result, it could be concluded that the petals in the hybrid nanoflowers were formed by regular arrangement of $Cu_3(PO_4)_2$ crystals and the inorganic composition of SF@Cu-NFs was $Cu_3(PO_4)_2$.

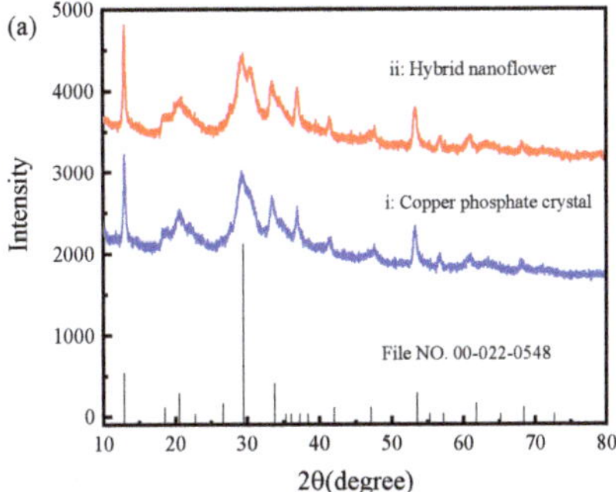

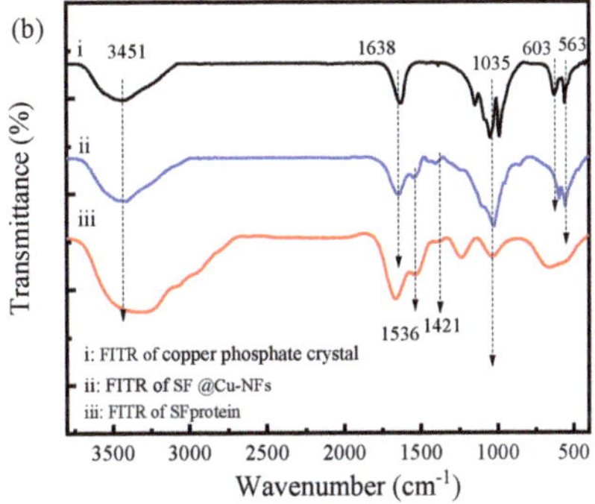

Figure 5. (**a**) XRD spectrum of (i) copper phosphate; (ii) SF@Cu-NFs nanoflower. (**b**) FTIR spectrum of (i) copper phosphate; (ii) SF@Cu-NFs nanoflower; (iii) SF protein.

By assigning peaks to various groups and bonds, the detail FTIR spectra of $Cu_3(PO_4)_2$ (spectrum i), SF@Cu-NFs product (spectrum ii) and SF protein (spectrum iii) were shown in Figure 5b. As it can be seen, SF@Cu-NFs showed (1) weak peaks at 563 cm^{-1} and 603 cm^{-1} corresponding to flexural vibration of P-O, (2) strong peaks at 1035 cm^{-1} corresponding to stretching vibration of P-O [59,60]. These signals corresponded to the asymmetric and symmetric stretching vibrations of PO_4^{3-}, which were also present in $Cu_3(PO_4)_2$ (spectrum i). Meanwhile, SF@Cu-NFs also showed (1) peak at 1638 cm^{-1} corresponding to stretching vibration of C-O originated from amide I, (2) peak at 1536 cm^{-1} corresponding to superposition of bending vibration of N-H and stretching vibration of C-N originated from amide II, (3) peak at 1421 cm^{-1} corresponding to bending vibration of N-H originated from amide III [61]. These signals corresponded to the major amide bands originating from SF protein, which were also present in SF protein (spectrum iii). Compared with $Cu_3(PO_4)_2$ and SF, SF@Cu-NFs showed no significant changes before and after the formation. Without new absorption peaks or obvious peak shifts, the above results indicated that SF protein was fixed by self-assembly rather than by covalent bonds. Meanwhile, the structural integrity of the silk fibroin protein remained intact after product formation, which furthermore identified the successful preparation of SF@Cu-NFs.

3.2.3. Component Analysis by TGA

In order to clarify the developed nanocomposites construction of the prepared product, TGA was used to demonstrate the existence of SF protein in SF@Cu-NFs based on the gravity measurement, shown as spectrum (i), (ii), and (iii) in Figure 6 and Table 1 for $Cu_3(PO_4)_2$, SF@Cu-NFs and SF protein, respectively. As can be see, although both the weight loss of $Cu_3(PO_4)_2$ and SF@Cu-NFs could be divided into three stages, but there were some differences between them. For $Cu_3(PO_4)_2$ as shown in spectrum (i), its weight loss included (1) due to the physical combination water; (2) due to part of chemical crystalline water; (3) due to rest of chemical crystalline water. However, the three weight loss stages of SF@Cu-NFs as shown in spectrum (ii) included (1) due to the physical combination water; (2) due to the chemical crystalline water; (3) due to the decomposition of amino acid residues and major peptide chains of SF protein. The weight loss of SF protein showed three stages, including (1) due to the physical combination water; (2) due to chemical crystalline water; (3) due to decomposition of amino acid residues and major peptide chains. The weight loss of chemical crystalline water showed the difference between $Cu_3(PO_4)_2$ and SF@Cu-NFs, which was mainly resulted from the different water content. This was mainly because $Cu_3(PO_4)_2$ was well-known to contain more crystalline waters like $CuSO_4 \cdot 5H_2O$ crystal but protein@Cu-NFs were generally considered to contain three crystalline waters in the form of $Cu_3(PO_4)_2 \cdot 3H_2O$ [31,39].

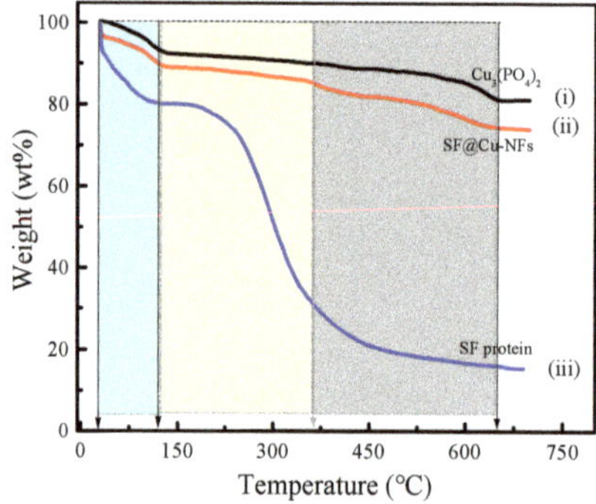

Figure 6. TGA spectrum of (i) Copper phosphate; (ii) SF@Cu-NFs nanoflower; (iii) SF protein.

Table 1. Thermogravimetric data of TGA spectrum (i) copper phosphate; (ii) SF@Cu-NFs nanoflower; (iii) SF protein.

Temperature Range	30–127 °C	127–361 °C	361–648 °C
$Cu_3(PO_4)_2$	7.32%	2.57%	8.67%
SF@Cu-NFs	10.25%	4.30%	10.85%
SF protein	19.71%	49.63%	14.48%

3.3. SF@Cu-NFs Adsorption Performance for Pb(II)

3.3.1. Investigation of pH Influence on Pb(II) Adsorption

The pH value is a very key factor for the investigation of HMIs adsorption, and the pH effect on the adsorption of Pb(II) by the prepared SF@Cu-NFs was investigated from the following two aspects in this work.

On one hand, pH can affect the existing form of the adsorbate metal ion. Generally speaking, HMI would produce hydrolysis with the change of pH and Pb(II) would exit with different forms at different pH (free ionic Pb^{2+} or solid $Pb(OH)_2$). In order to precisely study the adsorption process, Pb(II) should be excellently prevented from precipitating to be $Pb(OH)_2$. Consequently, the initial pH must be controlled for the accurate monitoring of Pb(II) adsorption. The pH can be calculated for Pb(II) by the following precipitation–dissolution equilibrium

$$Pb(OH)_2 \leftrightarrow Pb^{2+} + 2OH^-$$

(3)

Its equilibrium constant was

$$K_{sp(Pb(OH)_2)} = \left[Pb^{2+}\right]\left[OH^-\right]^2 \tag{4}$$

Then, *p*OH and *p*H were calculated to be

$$pOH = -lg\sqrt{\frac{K_{sp(Pb(OH)_2)}}{\left[Pb^{2+}\right]}} \tag{5}$$

and

$$pH = 14 + lg\sqrt{\frac{K_{sp(Pb(OH)_2)}}{\left[Pb^{2+}\right]}} \tag{6}$$

with $K_{sp(Pb(OH)_2)} = 1.2 \times 10^{-15}$ and trace amount $\left[Pb^{2+}\right] < 10^{-7}$ mol·L^{-1}, the *p*H for Pb(II) adsorption should be *p*H < 10 by Equation (6).

On the other hand, pH can affect the surface charge of the adsorbent material, which is a main factor influencing the adsorption capacity [62]. Then the zeta potentials of SF@Cu-NFs were analyzed at different pH conditions by calculating the average of 10 measurements. The test shown in Figure S3a indicates that the surface of the prepared SF@Cu-NFs adsorbent was positively charged at pH < 5 but negatively charged at pH > 5. The isoelectric point of SF@Cu-NFs was calculated to be pH = 4.2, at which the dispersion system of SF@Cu-NFs showed the lowest stability. Meanwhile, SF@Cu-NFs showed the highest stability with the highest absolute zeta potential of 12.83 mV at pH = 5.0, which was expected to have the best adsorption result.

Combinding the results of above two aspects, the influences of pH value on the adsorption were correspondingly investigated in the range of pH = 4.0–9.0 for 200 mg L^{-1} Pb(II) and the responses of adsorption capacity were shown in Figure S3b. It was noted that the adsorption capacities increased sharply with increasing pH value from 4.0 to 5.0, and then decreased slowly from 5.0 to to 9.0. As a result, the prepared SF@Cu-NFs showed the maximum adsorption capacity at pH = 5.0, which was consistent with above result of zeta potential investigation. Subsequently, pH = 5.0 was selected as the optimum condition to obtain the best Pb(II) adsorption.

3.3.2. Evaluation of Adsorption Kinetics

The kinetics investigation is important to choose the optimal operating condition on practical systems for HMIs removal [63]. To get a deeper understanding of the adsorption process of Pb(II) on SF@Cu-NFs, three typical kinetic models including pseudo-first-order (Equation (7)), pseudo-second-order (Equation (8)) and intraparticle diffusion (Equation (9)) models were used to analyze the experimental data

$$\ln(Q_e - Q_t) = \ln Q_e - k_1 t \tag{7}$$

$$\frac{t}{Q_t} = \frac{1}{k_2 Q_e^2} + \frac{t}{Q_e} \tag{8}$$

$$Q_t = C + k_n t^{0.5} \tag{9}$$

where t (min) is the adsorption time; Q_e and Q_t (mg g^{-1}) are the Pb(II) amount adsorbed at equilibrium; and k_1 (min^{-1}), k_2 (g mg^{-1} min^{-1}), and k_n (mg g^{-1} min$^{-1/2}$) are the rate constants of pseudo-first-order, pseudo-second-order, and intraparticle kinetics models, respectively. The adsorption capacity at different time was indicated in Figure 7a and the kinetic experimental data investigated by the three kinetic models were shown in Figure 7b–d, respectively. The fitting equations and kinetic parameters for the adsorption of Pb(II) by the prepared SF@Cu-NFs, as calculated from the plots of above three models, were listed in Table 1.

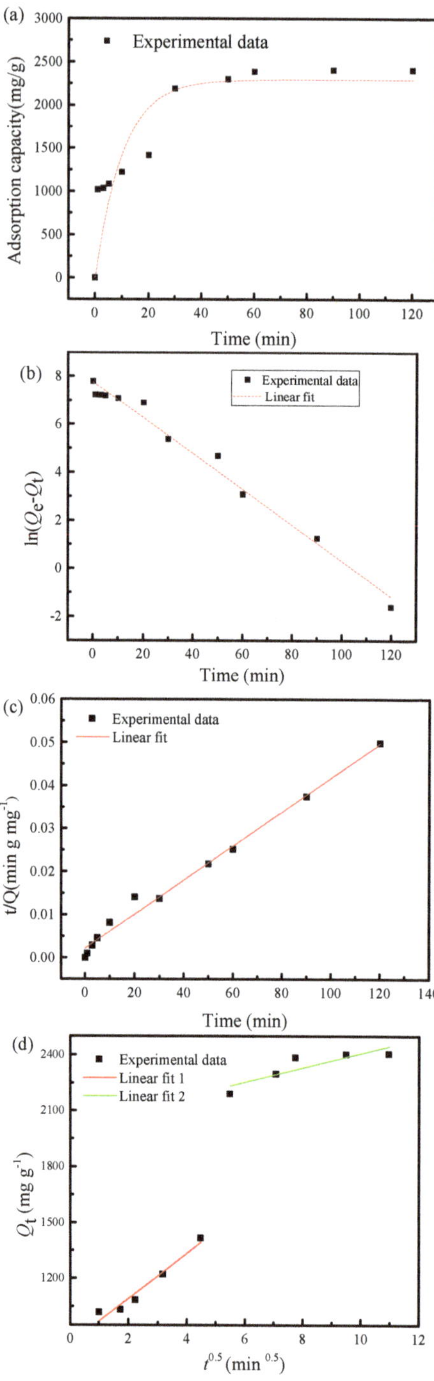

Figure 7. (**a**) The adsorption capacity at different times. Pb(II) removal by silk fibroin mediated nanoflowers (**b**) the pseudo-first-order; (**c**) the pseudo-second-order; (**d**) the intra-particle diffusion model sorption kinetics curves.

Pseudo-first-order (shown in Figure 7b) and pseudo-second-order (shown in Figure 7c) models were generally used to predict equilibrium adsorption capacity. With high correlation coefficient values of $R_2^2 = 0.99 > R_1^2 = 0.98$, the results indicated both the pseudo-second-order model provided a better fitting effect on the experimental data, demonstrating the calculated value of the pseudo-second-order model was closer to the actual value than that of pseudo-first-order model. As a result, the adsorption capacity of 300 mg L^{-1} Pb(II) was then estimated to be 2528.36 mg g^{-1} by the prepared SF@Cu-NFs. As shown in Table 2, the relative error between the calculated capacity of 2528.36 mg g^{-1} and experiment result of 2407.00 mg g^{-1} was evaluated to be 4.8%, which indicating the good agreement for them.

Table 2. Comparison parameters of the pseudo-first-order, the pseudo-second-order and the intra-particle diffusion models for Pb(II) adsorption by SF@Cu-NFs.

Model	Fitting Equation	Parameter	Value
Pseudo-first-order	$y = -0.07x + 7.75$	$Q_{e,cal1}$ (mg g^{-1})	2312.00
		k_1 (min^{-1})	0.075
		R_1^2	0.98
Pseudo-second-order	$y = 4 \times 10^{-4}x + 0.0023$	$Q_{e,cal2}$ (mg g^{-1})	2528.36
		k_2 (g mg^{-1} min^{-1})	6.92×10^{-5}
		R_2^2	0.99
Intra-particle diffusion	$y = 121.00x + 850.35$	C_1	850.35
		k_{ip1}	121.00
		R_{ip1}^2	0.94
	$y = 38.61x + 2022.19$	C_2	2022.19
		k_{ip2}	38.61
		R_{ip2}^2	0.70
Result in this work		$Q_{e,exp}$ (mg g^{-1})	2407.00

The intraparticle diffusion model was usually employed to examine the controlling mechanism such as transfer and chemical reaction for the adsorption process. As shown in Figure 7d, the fitting curve of intraparticle diffusion can be divided into two linear parts, indicating that the adsorption process consists of two steps. The first stage belongs to boundary layer adsorption, which is the diffusion of Pb(II) adsorbate from solution to SF@Cu-NFs surface. In the second stage, Pb(II) ion passes through the boundary layer to further react inside the SF@Cu-NFs adsorbent, which belongs to the intraparticle diffusion. Because the straight lines of the two stages do not pass through the origin of coordinate axis, it shows that the adsorption process is controlled by both intraparticle diffusion and boundary layer diffusion.

3.3.3. Adsorption Isotherm Experiment and Adsorption Thermodynamics

At the above optimal condition of pH = 5, the adsorption thermodynamics were further investigated by changing the initial Pb(II) concentrations with 1×10^{-3} g mL^{-1} SF@Cu-NFs adsorbent. The adsorption performances of Pb(II) on SF@Cu-NFs were studied by the following Langmuir (Equation (10)), Freundlich (Equation (11)) and Temkin (Equation (12)) models.

$$\frac{C_e}{Q_e} = \frac{C_e}{Q_{max}} + \frac{1}{K_L Q_{max}} \tag{10}$$

$$\ln Q_e = \ln K_F + \frac{1}{n} \ln C_e \tag{11}$$

$$Q_e = A \ln C_e + B \tag{12}$$

where K_L, K_F, and A are the equilibrium constants of Langmuir, Freundlich and Temkin adsorption, respectively; C_e is the equilibrium concentration of Pb(II); Q_e and Q_{max} are the amount of equilibrium adsorption capacity and the maximum adsorption capacity of Pb(II), respectively; The value of $n > 1$ suggests a normal Langmuir isotherm and $n < 1$ suggests the cooperative adsorption, respectively [64].

The adsorption of Pb(II) was investigated with different initial concentrations of 5–500 mg L^{-1} at different temperatures of 298 K. Meanwhile, due to the high adsorption ability of SF@Cu-NFs, almost all the Pb(II) in the solution were adsorbed completely in the low concentration of 5–50 mg L^{-1}. The free concentration of Pb(II) in the final solution could not be effectively detected and the equilibrium concentration of Pb(II) was thereafter regarded to be 0 during this concentration stage. As a result, the Langmuir, Freundlich, and Temkin adsorption isotherms of Pb(II) adsorption were efficiently fitted in high Pb(II) concentration of 80–500 mg L^{-1} and shown in Figure S4a–c, respectively.

For clarity, the fitting equations and parameters of Langmuir model, Freundlich model and Temkin model for Pb(II) by SF@Cu-NFs were summarized and listed in Table 3. It is found that Langmuir model provided better fitting to the equilibrium data than that of Freundlich and Temkin models with a higher correlation coefficient of 0.98, indicating that the adsorption of Pb(II) on the prepared SF@Cu-NFs belonged to monolayer adsorption instead of multilayer adsorption. Since the Langmuir model suggested that molecules are adsorbed uniformly, it can be deduced that the prepared SF@Cu-NFs were fairly homogeneous with SF protein assembly.

Table 3. Comparison parameters of Langmuir, Freundlich and Temkin models for Pb(II) adsorption by SF@Cu-NFs.

Fitting Model	Langmuir Model $y = 5.24 \times 10^{-4}x + 0.11$			Freundlich Model $y = 0.1895x + 6.36$			Temkin Model $y = 230.24x + 409.93$		
Parameter	Q_{max} (mg g^{-1})	K_L (mg L^{-1})	R^2	K_F (mg^{n-1} g^{-1} L^{-n})	n	R^2	A	B	R^2
Value	1908.39	4.76×10^{-3}	0.98	578.25	5.277	0.79	230.24	409.93	0.77

In order to evaluate the treat ability of the prepared SF@Cu-NFs for Pb(II) adsorption, the maximum adsorption capacity in this work was compared the results obtained by some other adsorbents which were reported previously. The results were listed in Table S2. As compared, SF@Cu-NFs indicated as an excellent adsorbent for Pb(II) treatment with the Q_{max} as high as 1908 mg g^{-1}, which was about 3–20 folds than that of the other adsorbents. As a result, SF@Cu-NFs was suggested to be a candidate for Pb(II) removal in wastewater with much higher adsorption performance.

In order to obtain the experimental parameters of adsorption thermodynamics on the Pb(II) adsorption by SF@Cu-NFs, the adsorption capacity and equilibrium constant at different temperatures were shown in Figure 8a,b, respectively. Then the thermodynamic data were calculated assuming the temperature-constant entropy and enthalpy of adsorption and according to the following temperature-related equations of Equation (13) to Equation (15).

$$\Delta_r G^{\theta}_{m(T)} = -RT \ln K^{\theta}_T \tag{13}$$

$$\Delta_r G^{\theta}_{m(T)} = \Delta_r H^{\theta}_m - T\Delta_r S^{\theta}_m \tag{14}$$

$$K^{\theta}_T = \frac{Q_e}{C_e} \tag{15}$$

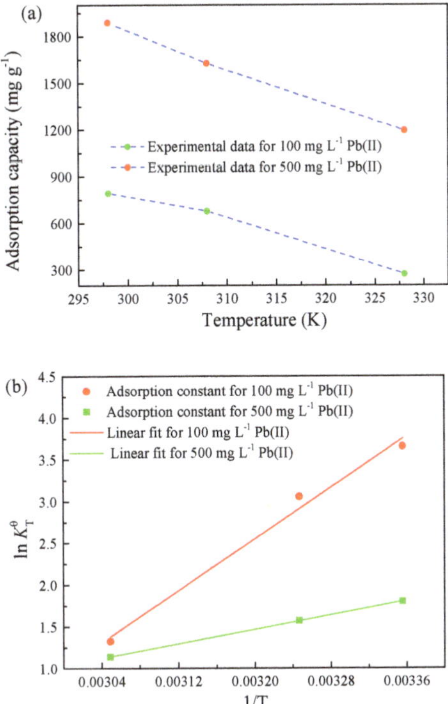

Figure 8. Effect of temperature on the equilibrium adsorption of Pb(II) (**a**) adsorption capacity and (**b**) adsorption constant.

The results of $\Delta_r G^{\theta}_{m(T)}$, $\Delta_r H^{\theta}_{m}$ and $\Delta_r S^{\theta}_{m}$ were listed in Table S3 and several conclusions could be obtained.

First, with $\Delta_r G^{\theta}_{m(T)} < 0$ for both 100 and 500 mg L^{-1} Pb(II), the adsorption process of Pb(II) onto SF@Cu-NFs was indicated to be spontaneous. However, the absolute value of $\Delta_r G^{\theta}_{m(T)}$ for Pb(II) adsorption was noted to decrease from 9.07 kJ mol^{-1} to 3.63 kJ mol^{-1} with the increase in temperature from 298 K to 328 K, which indicating that lower temperature was favored for the removal of Pb(II) by SF@Cu-NFs.

Second, with $\Delta_r H^{\theta}_{m} < 0$ for both 100 and 500 mg L^{-1} Pb(II), the adsorption process of Pb(II) onto SF@Cu-NFs was indicated to be exothermic. However, the absolute value of $\Delta_r H^{\theta}_{m}$ for Pb(II) adsorption was noted to decrease from 64.30 kJ mol^{-1} to 17.91 kJ mol^{-1} with the increase in Pb(II) concentration from 100 mg L^{-1} to 500 mg L^{-1}, which indicating that higher concentration would result in a mutual repulsion between mutual Pb(II).

Third, with the absolute value of $\Delta G < 40$ kJ mol^{-1} at different Pb(II) concentrations and different temperatures, the observations on the adsorption of Pb(II) by the SF@Cu-NFs in present study was an obvious physical adsorption process.

3.3.4. Investigation of Adsorption Selectivity for Pb(II)

Selectivity is one of the primary criteria for good adsorbents for the removal of trace amounts of heavy metals in the presence of other competing metal ions. In this work, the adsorption selectivity of SF@Cu-NFs was studied for three HMIs of Pb(II), Cd(II), and Ni(II) under the condition of 20 mL 100 mg L^{-1} HMIs with 3 mg SF@Cu-NFs adsorbent. The removal efficiency for different HMIs at different concentration were shown in Figure S5. For the three HMIs, all the adsorption processes were indicated to be rapid within the first 5 min and thereafter relatively slower by achieving the equilibrium in 90 min. The heavy metal ions adsorption efficiency (*AE*) can be calculated as

$$AE(\%) = \frac{(C_0 - C_e)}{C_0} \times 100\% \tag{16}$$

The adsorption efficiencies of Cd(II), Ni(II) and Pb(II) were calculated to be 23.77%, 18.76%, and 99.75%, indicating the much higher adsorption performance for Pb(II) by the prepared nanoflower. The selective factor (sf) was defined to evaluate the adsorbent selectivity as

$$sf = \frac{AE_a}{AE_b} \tag{17}$$

where AE_a and AE_b were adsorption efficiencies for the superior and inferior adsorption HMIs, respectively. For the prepared SF@Cu-NFs, its selective factors of Pb(II) were calculated to be 4.2 relative to Cd(II) and 5.3 relative to Ni(II), which proved the excellent adsorption selectivity for Pb(II) by SF@Cu-NFs.

3.4. SF@Cu-NFs Adsorption Mechanism for Pb(II)

3.4.1. Verification of Pb(II) Adsorption by SF@Cu-NFs

In order to access the interactions between SF@Cu-NFs adsorbent and Pb(II) ion, SF@Cu-NFs after Pb(II) adsorption was furthermore investigated by zeta potential, FTIR, and XRD measurements, which were respectively shown in Figure 9a,b and Figure S6.

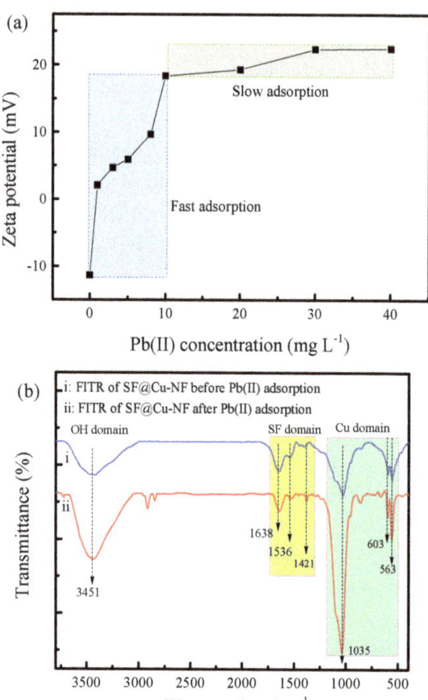

Figure 9. (**a**) Surface zeta potential measurement of SF@Cu-NFs at different Pb^{2+} concentrations; (**b**) FTIR spectra of SF@Cu-NFs before (spectrum i) and after (spectrum ii) Pb(II) adsorption.

The average zeta potentials of SF@Cu-NFs adsorbent were analyzed through DLS measurements. With the addition of different Pb(II) concentrations, the surface zeta potential measurement of SF@Cu-NFs was presented in Figure 9a. At the optimum pH = 5, the surface of the SF@Cu-NFs adsorbent was negatively charged and had an average zeta potential of about −12 mV without Pb(II). With the addition of Pb(II), an increase in the zeta potential was produced, which showed fast in lower Pb(II) concentration (indicated as the blue area) and thereafter varies rather slowly in higher Pb(II)concentration (indicated as the green area). This is a typical two-site adsorption behavior corresponding to two-type interaction dominance [65].

The pattern of HMI adsorption onto solid adsorbents can be attributable to the groups and bonds present on the material surface. In order to elucidate the active interaction site, FTIR spectrophotometry was performed to investigate the changes of functional groups of SF@Cu-NFs adsorbent before and after Pb(II) adsorption, which were shown as spectrum (i) and spectrum (ii) in Figure 9b, respectively. As it can be seen, the two FITR domains of SF@Cu-NFs after Pb(II) adsorption showed different groups and bonds, including (1) decreasing peaks at 1638, 1536, and 1421 cm^{-1} at SF domain, indicating the interaction between Pb(II) and functional N–H, C–O, and C–N groups; (2) increasing peaks at 1035, 603, and 563 cm^{-1} at Cu domain, indicating the interaction between Pb(II) and P-O groups. Because there was no new functional group appearing in the SF@Cu-NFs adsorbent after Pb(II) adsorption, it can be determined that the interaction between Pb(II) and SF@Cu-NFs belongs to physical but not chemical adsorption. The two bands around 2800 cm^{-1} are corresponding to stretching vibration of saturated C-H. Generally, the group with high electronegativity has strong ability of electron absorption. When it is connected with the number of carbon atoms on the carbonyl group of alkyl ketone, the electron cloud will shift from oxygen atom to the middle of double bond due to the induction effect. These increase the force constant of C=O bond, increases the vibration frequency of C=O, and shifts the absorption peak to a higher wave number. This result is also consistent to the adsorption energy obtained in thermodynamic investigation, which was calculated to be $\Delta G < 40$ kJ mol^{-1}.

As presented in Figure S6, XRD patterns of SF@Cu-NFs were measured before and after Pb(II) uptake. Compared to the diffraction peaks of hybrid nanoflowers before adsorbing Pb(II), there were new several miscellaneous diffraction peaks at 2θ values of 21.5, 26.2, 27.5, 30.0, which confirmed hybrid nanoflowers successfully adsorbed heavy metal ion Pb(II).

3.4.2. Mechanism Analysis of Pb(II) Adsorption by SF@Cu-NFs

Based on the results of adsorption property and adsorption characterization mentioned above, the mechanism is proposed to illustrate the elimination performance for Pb(II) by the prepared SF@Cu-NFs. The mechanism diagram is schematically presented in Figure 10, in which the chemical structure of SF protein was referenced from the previous report [66] and the electronic structure of Cu$_3$(PO$_4$)$_2$·3H$_2$O was calculated by Material Studio 7.0. Blue, yellow, red, and grey spheres designate Cu, P, O, and H atoms, respectively. The adsorption of SF@Cu-NFs for Pb(II) removal was originated from two types of adsorption sites and two kinds of interaction dominance, which can be ascribed to the individual organic SF protein and inorganic Cu$_3$(PO$_4$)$_2$ crystal. Correspondingly, two stages of fast adsorption and slow adsorption of Pb(II) by the prepared SF@Cu-NFs was revealed and described as follows.

Fast adsorption stage of Pb(II). For this stage, the flower 'stamen' of organic SF protein was designated as responsible adsorption site for fast adsorption of Pb(II) (shown as upper part in Figure 10). This kind adsorption was originated from multiple coordinative interaction produced between Pb(II) and abundant N, O elements. This interaction showed strong due to the numerous amide groups provided by SF protein. Meanwhile, the fast adsorption occurred in the shorter adsorption time (shown as the first linear part by intraparticle kinetic investigation in Figure 7c) and in the lower adsorbent concentration (shown as the first increasing part by zeta potential measurement in Figure 9a).

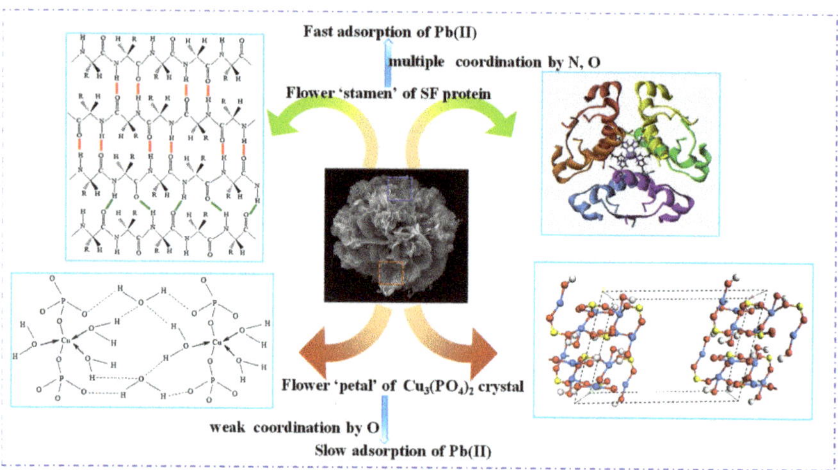

Figure 10. Proposed adsorption mechanism of Pb(II) by SF@Cu-NFs. Upper part indicates the fast adsorption of Pb(II) by the organic SF component and lower part indicates the slow adsorption of Pb(II) by the inorganic $Cu_3(PO_4)_2$ component.

Slow adsorption stage of Pb(II). For this stage, the flower 'petal' of inorganic $Cu_3(PO_4)_2$ crystal was designated as responsible adsorption site for slow adsorption of Pb(II) (shown as lower part in Figure 10). This kind adsorption was originated from unique coordinative interaction produced between Pb(II) and O element. This interaction showed weak due to the powerful restriction from the strong ion bond by Cu(II) elements in $Cu_3(PO_4)_2$ crystal. Meanwhile, the slow adsorption occurred in the longer adsorption time (shown as the second linear part by intraparticle kinetic investigation in Figure 7c) and in the higher adsorbent concentration (shown as the second increasing part by zeta potential measurement in Figure 9a).

4. Conclusions

In this work, natural material of SF protein was used for the fabrication of protein–inorganic hybrid nanoflowers through self-assembly and the three-dimensional structure was applied to efficient adsorption of HMI Pb(II).

Through adsorption isotherms and kinetics, the adsorption performance of SF@Cu-HNFs for Pb(II) removal was systematically evaluated in detail. Langmuir and pseudo-second-order models indicated the monolayer adsorption and high capacity on the SF@Cu-NFs. Meanwhile, the adsorption thermodynamics showed that the spontaneous and exothermic process. As compared, SF@Cu-NFs indicated as an excellent adsorbent for Pb(II) treatment with the Q_{max} as high as 1908 mg g^{-1}, which was about 3–20 folds greater than that of the other adsorbents.

By ascribing to its individual organic and inorganic component, the adsorption mechanism of SF@Cu-NFs for Pb(II) removal was discussed and revealed with two stages of fast adsorption and slow adsorption. On one hand, the flower 'stamen' of organic SF protein was designated as responsible adsorption site for fast adsorption of Pb(II). On the other hand, the flower 'petal' of inorganic $Cu_3(PO_4)_2$ crystal was designated as responsible adsorption site for slow adsorption of Pb(II). This result clearly indicated that the silk fibroin protein-derived hybrid nanoflower could adsorb HMI Pb(II) well because of the adsorption site on the adsorbent surface.

In this work, we further understand the adsorption behavior and interaction process of HMI Pb(II) on the surface of silk fibroin derived hybrid nanoflowers. The present study has been successful in revealing the microscopic interaction process of Pb (II) adsorption that provides a new insight on understanding the adsorption mechanism. Also, based on interfacial adsorption, it is of great significance to comprehend the development of heavy metal ion removal applications. By fabricating SF@Cu-HNFs hybrid nanoflowers derived from SF protein, this work not only successfully provides insights on its adsorption performance and interaction mechanism for Pb(II) removal, but also significantly indicates its potential applications in contamination adsorption for environmental treatment.

Supplementary Materials: The following are available online at http://www.mdpi.com/1996-1944/13/5/1241/s1, Figure S1: Photography of solution change at reaction times of 0 min and 10 min., Figure S2: Diameter distribution of products from 0 to 24 h, Figure S3: (a) The surface zeta potential measurement of SF@Cu-NFs at pH=4.0–9.0; (b) Adsorption capacity of the prepared SF@Cu-NFs for Pb(II) at pH=4.0–9.0, Figure S4: The adsorption isotherms of Pb(II) fitting by (a) Langmuir model, (b) Freundlich model and (c)Temkin model, Figure S5: Evaluation of SF@Cu-NFs adsorption selectivity for (i) Pb(II), (ii) Cd(II) and (iii) Ni(II). Insets are HMIs solution after different adsorption of 0, 5, 20, 40 and 90 min. The chromogenic reagent for Pb(II) and Cd(II) are DTZ and for Ni(II) is DMG, Figure S6: XRD spectra of SF@Cu-NFs before (spectrum i) and after (spectrum ii) Pb(II) adsorption, Table S1: The related EDS data of SF@Cu-NFs, Table S2: Comparison maximum capacity of SF@Cu-NFs products in this work and other adsorbents for Pb(II) adsorption., Table S3: Thermodynamic data for Pb(II) adsorption by SF@Cu-NFs at different temperatures.

Author Contributions: M.D., X.L. and Y.X. designed experiments; J.L., C.Z., S.Q. and X.L. carried out experiments; H.W., S.F., and R.Z. analyzed experimental results. X.L. and Y.X. analyzed sequencing data. X.L. and Y.X. wrote the manuscript. All authors have read and agreed to the published version of the manuscript.

Funding: This research was funded by [the National Natural Science Foundation of China] grant number [51974266 and 51404203], [the China Postdoctoral Science Foundation Funded Project] grant number [2017M612993] and [the Miaozi Project of Scientific and Technological Innovation of Sichuan Province] grant number [2019091 and 2019093]. The APC was funded by [2019091 and 2019093].

Conflicts of Interest: The authors declare no competing financial interest.

References

1. Gumpu, M.B.; Sethuraman, S.; Krishnan, U.M.; Rayappan, J.B.B. A review on detection of heavy metal ions in water-An electrochemical approach. *Sens. Actuators B Chem.* **2015**, *213*, 515–533. [CrossRef]

2. Aragay, G.; Pons, J.; Merkoci, A. Recent trends in macro-, micro-, and nanomaterial-based tools and strategies for heavy-metal detection. *Chem. Rev.* **2011**, *111*, 3433–3458. [CrossRef]

3. He, Z.L.; Yang, X.E.; Stoffella, P.J. Trace elements in agroecosystems and impacts on the environment. *J. Trace Elem. Med. Biol.* **2005**, *19*, 125–140. [CrossRef]

4. Needleman, H. Lead poisoning. *Annu. Rev. Med.* **2004**, *55*, 209–222. [CrossRef]

5. Ilbeigi, V.; Valadbeigi, Y.; Tabrizchi, M. Ion Mobility Spectrometry of Heavy Metals. *Anal. Chem.* **2016**, *88*, 7324–7328. [CrossRef] [PubMed]

6. Patrick, L. Lead toxicity, a review of the literature. Part 1: Exposure, evaluation, and treatment. *Altern. Med. Rev.* **2006**, *11*, 2–22. [PubMed]

7. Moyo, M.; Guyo, U.; Mawenyiyo, G.; Zinyama, N.P.; Nyamunda, B.C. Marula seed husk (Sclerocarya birrea) biomass as a low cost biosorbent for removal of Pb(II) and Cu(II) from aqueous solution. *J. Ind. Eng. Chem.* **2015**, *27*, 126–132. [CrossRef]

8. Li, Y.; Song, S.; Xia, L.; Yin, H.; García Meza, J.V.; Ju, W. Enhanced Pb(II) removal by algal-based biosorbent cultivated in high-phosphorus cultures. *Chem. Eng. J.* **2019**, *361*, 167–179. [CrossRef]

9. Fu, F.; Wang, Q. Removal of heavy metal ions from wastewaters: A review. *J. Environ. Manag.* **2011**, *92*, 407–418. [CrossRef]

10. Gupta, V.K.; Ali, I.; Saleh, T.A.; Nayak, A.; Agarwal, S. Chemical treatment technologies for waste-water recycling-an overview. *RSC Adv.* **2012**, *2*, 6380–6388. [CrossRef]

11. Hao, T.-w.; Xiang, P.-y.; Mackey, H.R.; Chi, K.; Lu, H.; Chui, H.-K.; van Loosdrecht, M.C.M.; Chen, G.-H. A review of biological sulfate conversions in wastewater treatment. *Water Res.* **2014**, *65*, 1–21. [CrossRef] [PubMed]

12. Abdi, G.; Alizadeh, A.; Zinadini, S.; Moradi, G. Removal of dye and heavy metal ion using a novel synthetic polyethersulfone nanofiltration membrane modified by magnetic graphene oxide/metformin hybrid. *J. Membr. Sci.* **2018**, *552*, 326–335. [CrossRef]

13. Bolisetty, S.; Mezzenga, R. Amyloid-carbon hybrid membranes for universal water purification. *Nat. Nanotechnol.* **2016**, *11*, 365–371. [CrossRef] [PubMed]

14. Huang, D.; Wu, J.; Wang, L.; Liu, X.; Meng, J.; Tang, X.; Tang, C.; Xu, J. Novel insight into adsorption and co-adsorption of heavy metal ions and an organic pollutant by magnetic graphene nanomaterials in water. *Chem. Eng. J.* **2019**, *358*, 1399–1409. [CrossRef]

15. Yuan, Q.; Li, P.; Liu, J.; Lin, Y.; Cai, Y.; Ye, Y.; Liang, C. Facet-Dependent Selective Adsorption of Mn-Doped α-Fe$_2$O$_3$ Nanocrystals toward Heavy-Metal Ions. *Chem. Mater.* **2017**, *29*, 10198–10205. [CrossRef]

16. Lagadic, I.L.; Mitchell, M.K.; Payne, B.D. Highly Effective Adsorption of Heavy Metal Ions by a Thiol-Functionalized Magnesium Phyllosilicate Clay. *Environ. Sci. Technol.* **2001**, *35*, 984–990. [CrossRef]

17. Cao, C.-Y.; Qu, J.; Yan, W.-S.; Zhu, J.-F.; Wu, Z.-Y.; Song, W.-G. Low-Cost Synthesis of Flowerlike α-Fe$_2$O$_3$ Nanostructures for Heavy Metal Ion Removal: Adsorption Property and Mechanism. *Langmuir* **2012**, *28*, 4573–4579. [CrossRef] [PubMed]

18. Ide, Y.; Ochi, N.; Ogawa, M. Effective and Selective Adsorption of Zn^{2+} from Seawater on a Layered Silicate. *Angew. Chem. Int. Ed.* **2011**, *50*, 654–656. [CrossRef]

19. Ge, Y.; Li, Z. Application of Lignin and Its Derivatives in Adsorption of Heavy Metal Ions in Water: A Review. *ACS Sustain. Chem. Eng.* **2018**, *6*, 7181–7192. [CrossRef]

20. Chen, Q.; Zheng, J.; Zheng, L.; Dang, Z.; Zhang, L. Classical theory and electron-scale view of exceptional Cd(II) adsorption onto mesoporous cellulose biochar via experimental analysis coupled with DFT calculations. *Chem. Eng. J.* **2018**, *350*, 1000–1009. [CrossRef]

21. Sellaoui, L.; Soetaredjo, F.E.; Ismadji, S.; Bonilla-Petriciolet, A.; Belver, C.; Bedia, J.; Ben Lamine, A.; Erto, A. Insights on the statistical physics modeling of the adsorption of Cd^{2+} and Pb^{2+} ions on bentonite-chitosan composite in single and binary systems. *Chem. Eng. J.* **2018**, *354*, 569–576. [CrossRef]

22. Ma, J.; Liu, Y.; Ali, O.; Wei, Y.; Zhang, S.; Zhang, Y.; Cai, T.; Liu, C.; Luo, S. Fast adsorption of heavy metal ions by waste cotton fabrics based double network hydrogel and influencing factors insight. *J. Hazard. Mater.* **2018**, *344*, 1034–1042. [CrossRef] [PubMed]

23. Shi, Z.; Di Toro, D.M.; Allen, H.E.; Sparks, D.L. A general model for kinetics of heavy metal adsorption and desorption on soils. *Environ. Sci. Technol.* **2013**, *47*, 3761–3767. [CrossRef] [PubMed]

24. Efome, J.E.; Rana, D.; Matsuura, T.; Lan, C.Q. Insight Studies on Metal-Organic Framework Nanofibrous Membrane Adsorption and Activation for Heavy Metal Ions Removal from Aqueous Solution. *ACS Appl. Mater. Interfaces* **2018**, *10*, 18619–18629. [CrossRef] [PubMed]

25. Zhang, C.-Z.; Yuan, Y.; Guo, Z. Experimental study on functional graphene oxide containing many primary amino groups fast-adsorbing heavy metal ions and adsorption mechanism. *Sep. Sci. Technol.* **2018**, *53*, 1666–1677. [CrossRef]

26. Kampalanonwat, P.; Supaphol, P. Preparation and Adsorption Behavior of Aminated Electrospun Polyacrylonitrile Nanofiber Mats for Heavy Metal Ion Removal. *ACS Appl. Mater. Interfaces* **2010**, *2*, 3619–3627. [CrossRef]

27. Kyzas, G.Z.; Bomis, G.; Kosheleva, R.I.; Efthimiadou, E.K.; Favvasa, E.P.; Kostoglou, M.; Mitropoulos, A.C. Nanobubbles effect on heavy metal ions adsorption by activated carbon. *Chem. Eng. J.* **2019**, *356*, 91–97. [CrossRef]

28. Cao, C.-Y.; Wei, F.; Qu, J.; Song, W.-G. Programmed Synthesis of Magnetic Magnesium Silicate Nanotubes with High Adsorption Capacities for Lead and Cadmium Ions. *Chem. A Eur. J.* **2013**, *19*, 1558–1562. [CrossRef]

29. Negron, L.M.; Diaz, T.L.; Ortiz-Quiles, E.O.; Dieppa-Matos, D.; Madera-Soto, B.; Rivera, J.M. Organic Nanoflowers from a Wide Variety of Molecules Templated by a Hierarchical Supramolecular Scaffold. *Langmuir* **2016**, *32*, 2283–2290. [CrossRef]

30. Jiang, D.; Yang, Y.; Huang, C.; Huang, M.; Chen, J.; Rao, T.; Ran, X. Removal of the heavy metal ion nickel (II) via an adsorption method using flower globular magnesium hydroxide. *J. Hazard. Mater.* **2019**, *373*, 131–140. [CrossRef]

31. Ge, J.; Lei, J.; Zare, R.N. Protein-inorganic hybrid nanoflowers. *Nat. Nanotechnol.* **2012**, *7*, 428–432. [CrossRef] [PubMed]

32. Cao, H.; Yang, D.-P.; Ye, D.; Zhang, X.; Fang, X.; Zhang, S.; Liu, B.; Kong, J. Protein-inorganic hybrid nanoflowers as ultrasensitive electrochemical cytosensing Interfaces for evaluation of cell surface sialic acid. *Biosens. Bioelectron.* **2015**, *68*, 329–335. [CrossRef] [PubMed]

33. Zhu, X.; Huang, J.; Liu, J.; Zhang, H.; Jiang, J.; Yu, R. A dual enzyme-inorganic hybrid nanoflower incorporated microfluidic paper-based analytic device (mu PAD) biosensor for sensitive visualized detection of glucose. *Nanoscale* **2017**, *9*, 5658–5663. [CrossRef] [PubMed]

34. Zhu, J.; Wen, M.; Wen, W.; Du, D.; Zhang, X.; Wang, S.; Lin, Y. Recent progress in biosensors based on organic–inorganic hybrid nanoflowers. *Biosens. Bioelectron.* **2018**, *120*, 175–187. [CrossRef]

35. Thawari, A.G.; Rao, C. Peroxidase-like Catalytic Activity of Copper-Mediated Protein-Inorganic Hybrid Nanoflowers and Nanofibers of beta-Lactoglobulin and alpha-Lactalbumin: Synthesis, Spectral Characterization, Microscopic Features, and Catalytic Activity. *ACS Appl. Mater. Interfaces* **2016**, *8*, 10392–10402. [CrossRef]

36. Altinkaynak, C.; Tavlasoglu, S.; Ozdemir, N.; Ocsoy, I. A new generation approach in enzyme immobilization: Organic–inorganic hybrid nanoflowers with enhanced catalytic activity and stability. *Enzym. Microb. Technol.* **2016**, *93–94*, 105–112. [CrossRef]

37. Zheng, L.; Sun, Y.; Wang, J.; Huang, H.; Geng, X.; Tong, Y.; Wang, Z. Preparation of a Flower-Like Immobilized D-Psicose 3-Epimerase with Enhanced Catalytic Performance. *Catalysts* **2018**, *8*, 468. [CrossRef]

38. Chen, W.; Tian, R.; Xu, C.; Yung, B.C.; Wang, G.; Liu, Y.; Ni, Q.; Zhang, F.; Zhou, Z.; Wang, J.; et al. Microneedle-array patches loaded with dual mineralized protein/peptide particles for type 2 diabetes therapy. *Nature Commun.* **2017**, *8*, 1–11. [CrossRef]

39. Sun, J.; Ge, J.; Liu, W.; Lan, M.; Zhang, H.; Wang, P.; Wang, Y.; Niu, Z. Multi-enzyme co-embedded organic–inorganic hybrid nanoflowers: Synthesis and application as a colorimetric sensor. *Nanoscale* **2014**, *6*, 255–262. [CrossRef]

40. Zhang, Z.; Zhang, Y.; Song, R.; Wang, M.; Yan, F.; He, L.; Feng, X.; Fang, S.; Zhao, J.; Zhang, H. Manganese(II) phosphate nanoflowers as electrochemical biosensorsfor the high-sensitivity detection of ractopamine. *Sens. Actuators B Chem.* **2015**, *211*, 310–317. [CrossRef]

41. Omenetto, F.G.; Kaplan, D.L. New Opportunities for an Ancient Material. *Science* **2010**, *329*, 528–531. [CrossRef]

42. Jin, H.J.; Fridrikh, S.V.; Rutledge, G.C.; Kaplan, D.L. Electrospinning Bombyx mori Silk with Poly(ethylene oxide). *Biomacromolecules* **2002**, *3*, 1233–1239. [CrossRef] [PubMed]

43. Yang, Y.; Chen, X.; Ding, F.; Zhang, P.; Liu, J.; Gu, X. Biocompatibility evaluation of silk fibroin with peripheral nerve tissues and cells in vitro. *Biomaterials* **2007**, *28*, 1643–1652. [CrossRef] [PubMed]

44. Vepari, C.; Kaplan, D.L. Silk as a Biomaterial. *Prog. Polym. Sci.* **2007**, *32*, 991–1007. [CrossRef] [PubMed]

45. Kook, G.; Jeong, S.; Kim, S.H.; Kim, M.K.; Lee, S.; Cho, I.-J.; Choi, N.; Lee, H.J. Wafer-Scale Multilayer Fabrication for Silk Fibroin-Based Microelectronics. *ACS Appl. Mater. Interfaces* **2019**, *11*, 115–124. [CrossRef] [PubMed]

46. Amsden, J.J.; Perry, H.; Boriskina, S.V.; Gopinath, A.; Kaplan, D.L.; Dal, L.N.; Omenetto, F.G. Spectral analysis of induced color change on periodically nanopatterned silk films. *Opt. Express* **2009**, *17*, 21271–21279. [CrossRef] [PubMed]

47. Wang, J.; Yang, S.; Li, C.; Miao, Y.; Zhu, L.; Mao, C.; Yang, M. Nucleation and Assembly of Silica into Protein-Based Nanocomposites as Effective Anticancer Drug Carriers Using Self-Assembled Silk Protein Nanostructures as Biotemplates. *ACS Appl. Mater. Interfaces* **2017**, *9*, 22259–22267. [CrossRef]

48. Wang, X.; Yucel, T.; Lu, Q.; Hu, X.; Kaplan, D.L. Silk nanospheres and microspheres from silk/pva blend films for drug delivery. *Biomaterials* **2010**, *31*, 1025–1035. [CrossRef]

49. Yi, S.; Dai, F.; Ma, Y.; Yan, T.; Si, Y.; Sun, G. Ultrafine Silk-Derived Nanofibrous Membranes Exhibiting Effective Lysozyme Adsorption. *ACS Sustain. Chem. Eng.* **2017**, *5*, 8777–8784. [CrossRef]

50. Lei, Z.; Gao, C.; Chen, L.; He, Y.; Ma, W.; Lin, Z. Recent advances in biomolecule immobilization based on self-assembly: Organic–inorganic hybrid nanoflowers and metal-organic frameworks as novel substrates. *J. Mater. Chem. B* **2018**, *6*, 1581–1594. [CrossRef]

51. Cooper, J.W.; Chen, J.; Li, Y.; Lee, C.S. Membrane-based nanoscale proteolytic reactor enabling protein digestion, peptide separation, and protein identification using mass spectrometry. *Anal. Chem.* **2003**, *75*, 1067–1074. [CrossRef] [PubMed]

52. Sheldon, R. Enzyme Immobilization: The Quest for Optimum Performance. *Adv. Synth. Catal.* **2007**, *349*, 1289–1307. [CrossRef]

53. Brenckle, M.A.; Partlow, B.; Tao, H.; Applegate, M.B.; Reeves, A.; Paquette, M.; Marelli, B.; Kaplan, D.L.; Omenetto, F.G. Methods and Applications of Multilayer Silk Fibroin Laminates Based on Spatially Controlled Welding in Protein Films. *Adv. Funct. Mater.* **2016**, *26*, 44–50. [CrossRef]

54. Brenckle, M.A.; Tao, H.; Kim, S.; Paquette, M.; Kaplan, D.L.; Omenetto, F.G. Protein-Protein Nanoimprinting of Silk Fibroin Films. *Adv. Mater.* **2013**, *25*, 2409–2414. [CrossRef] [PubMed]

55. Shen, L.; Bao, N.; Prevelige, P.E.; Gupta, A. Fabrication of Ordered Nanostructures of Sulfide Nanocrystal Assemblies over Self-Assembled Genetically Engineered P22 Coat Protein. *J. Am. Chem. Soc.* **2010**, *132*, 17354–17357. [CrossRef]

56. Patel, S.K.S.; Otari, S.V.; Li, J.; Kim, D.R.; Kim, S.C.; Cho, B.-K.; Kalia, V.C.; Kang, Y.C.; Lee, J.-K. Synthesis of cross-linked protein-metal hybrid nanoflowers and its application in repeated batch decolorization of synthetic dyes. *J. Hazard. Mater.* **2018**, *347*, 442–450. [CrossRef]

57. Lin, Z.; Xiao, Y.; Yin, Y.; Hu, W.; Liu, W.; Yang, H. Facile Synthesis of Enzyme-Inorganic Hybrid Nanoflowers and Its Application as a Colorimetric Platform for Visual Detection of Hydrogen Peroxide and Phenol. *ACS Appl. Mater. Interfaces* **2014**, *6*, 10775–10782. [CrossRef]

58. Jianhua, H.; Chuanbao, C.; Faryal, I.; Xilan, M. Hierarchical porous nitrogen-doped carbon nanosheets derived from silk for ultrahigh-capacity battery anodes and supercapacitors. *ACS Nano* **2015**, *9*, 2556–2564.

59. Zhang, Z.; Zhang, Y.; He, L.; Yang, Y.; Liu, S.; Wang, M.; Fang, S.; Fu, G. A feasible synthesis of $Mn_3(PO_4)_2$@BSA nanoflowers and its application as the support nanomaterial for Pt catalyst. *J. Power Sources* **2015**, *284*, 170–177. [CrossRef]

60. Zhao, X.Y.; Zhu, Y.J.; Zhao, J.; Lu, B.Q.; Chen, F.; Qi, C.; Wu, J. Hydroxyapatite nanosheet-assembled microspheres: Hemoglobin-templated synthesis and adsorption for heavy metal ions. *J. Colloid. Interface Sci.* **2014**, *416*, 11–18. [CrossRef]

61. Aziz, S.; Sabzi, M.; Fattahi, A.; Arkan, E. Electrospun silk fibroin/PAN double-layer nanofibrous membranes containing polyaniline/TiO2 nanoparticles for anionic dye removal. *J. Polym. Res.* **2017**, *24*. [CrossRef]

62. Gurbuz, O.; Okutan, M. Structural, electrical, and dielectric properties of Cr doped ZnO thin films: Role of Cr concentration. *Appl. Surf. Sci.* **2016**, *387*, 1211–1218. [CrossRef]

63. Yuan, Y.; Zhang, L.; Xing, J.; Utama, M.I.B.; Lu, X.; Du, K.; Li, Y.; Hu, X.; Wang, S.; Genc, A.; et al. High-yield synthesis and optical properties of g-C3N4. *Nanoscale* **2015**, *7*, 12343–12350. [CrossRef] [PubMed]

64. Roosta, M.; Ghaedi, M.; Shokri, N.; Daneshfar, A.; Sahraei, R.; Asghari, A. Optimization of the combined ultrasonic assisted/adsorption method for the removal of malachite green by gold nanoparticles loaded on activated carbon: Experimental design. *Spectrochim. Acta Part A Mol. Biomol. Spectrosc.* **2014**, *118*, 55–65. [CrossRef] [PubMed]

65. Dong, Y.; Pappu, S.V.; Xu, Z. Detection of Local Density Distribution of Isolated Silanol Groups on Planar Silica Surfaces Using Nonlinear Optical Molecular Probes. *Anal. Chem.* **1998**, *70*, 4730–4735. [CrossRef]

66. Tetsuo, A.; Akio, K.; Ryoko, T.; Saitô, H. Conformational characterization of Bombyx mori silk fibroin in the solid state by high-frequency carbon-13 cross polarization-magic angle spinning NMR, x-ray diffraction, and infrared spectroscopy. *Macromolecules* **1985**, *18*, 1841–1845.

MDPI

St. Alban-Anlage 66

4052 Basel

Switzerland

Tel. +41 61 683 77 34

Fax +41 61 302 89 18

www.mdpi.com

Materials Editorial Office

E-mail: materials@mdpi.com

www.mdpi.com/journal/materials